Strategic Management of IT in Construction

Strategic Management of IT in Construction

Edited by

Martin Betts
Professor of Construction IT
Head of School of Construction and Property Management
University of Salford

Editorial Offices:
Osney Mead, Oxford OX2 0EL
25 John Street, London WC1N 2BL
23 Ainslie Place, Edinburgh EH3 6AJ
350 Main Street, Malden
MA 02148 5018, USA
54 University Street, Carlton
Victoria 3053, Australia
10, rue Casimir Delavigne
75006 Paris, France

Other Editorial Offices:

Blackwell Wissenschafts-Verlag GmbH
Kurfürstendamm 57
10707 Berlin, Germany

Blackwell Science KK
MG Kodenmacho Building
7–10 Kodenmacho Nihombashi
Chuo-ku, Tokyo 104, Japan

First published 1999

Set in 10/13 Palatino
by DP Photosetting, Aylesbury, Bucks
Printed and bound in Great Britain by
MPG Books Ltd, Bodmin, Cornwall

DISTRIBUTORS

Marston Book Services Ltd
PO Box 269
Abingdon
Oxon OX14 4YN
(*Orders:* Tel: 01235 465500
Fax: 01235 465555)

USA
Blackwell Science, Inc.
Commerce Place
350 Main Street
Malden, MA 02148 5018
(*Orders:* Tel: 800 759 6102
781 388 8250
Fax: 781 388 8255)

Canada
Login Brothers Book Company
324 Saulteaux Crescent
Winnipeg, Manitoba R3J 3T2
(*Orders:* Tel: 204 837-2987
Fax: 204 837-3116)

Australia
Blackwell Science Pty Ltd
54 University Street
Carlton, Victoria 3053
(*Orders:* Tel: 03 9347 0300
Fax: 03 9347 5001)

A catalogue record for this title
is available from the British Library

ISBN 0-632-04026-2

Library of Congress
Cataloging-in-Publication Data
Strategic management of IT in construction/edited
by Martin Betts.
p. cm.
Includes bibliographical references and index.
ISBN 0-632-04026-2
1. Construction industry – Communication systems. 2. Information technology. I. Title: Strategic management of information technology in construction.
TH215.S77 1999
624'.068'4 – dc21 99-31142
CIP

For further information on Blackwell Science, visit our website: www.blackwell-science.com

Contents

Foreword

Books are sometimes referred to as timely. There can be little doubt that in this volume of writings, the construction industry is provided with a timely insight into two of the major factors which will lead to major change in the construction industry in the early part of the new millennium. These factors are, first, construction process re-engineering and, second, the information systems which will support such change.

For centuries, the structure of the construction industry, together with its practices and techniques, have evolved through informal and, more recently, formal means. Apprenticeship and education of one form or another were the dominant forms of skill development in construction. Skills have been passed on and improvements encouraged by modification and enhancement, but at root the industry remains craft based. There is very little rigorous knowledge capture or analysis, often with the excuse that 'the industry is different'. For those who come into construction from outside, this excuse can appear weak and lead to unwarranted complacency in an increasingly competitive world. They would argue that many issues of construction are common to all process-based industries and there is much to learn from what has been achieved elsewhere. There is a need to shift from the 'art' of construction to the 'science' of construction. This involves measurement, data capture and modelling that allow repetition of good practice beyond the inherited memory of good professionals. Unless we create this body of knowledge in a systematic and rigorous manner we cannot define where we come from or where we are going to improve. The tools for this task are now available and are being used by research groups working with practitioners as exemplified in this volume.

Often these tools require major processing power beyond manual methods. With the advent of personal computers over the past 20 years, this power is now available on every work desk. The information systems within these machines provide new opportunities for change and improvement that can transform the construction industry. Salford University has identified the major themes within these systems for industry development as **visualisation**, **intelligence**, **construction** and **integration**, and has structured its research programme accordingly. Of these it is the integration theme which is paramount and is aided by the others. This theme in itself allows the fragmented processes of construction to be brought together and the total process to be analysed and addressed in a more systematic way than what has been achieved in the past.

These major levers for change do not ignore the important developments of the past or the techniques that are used so extensively and to such major benefit in the industry today. Rather they harness them in a structured way so that the whole becomes more than the sum of the parts and tests their efficiency in producing quality buildings on time and to cost.

This book captures some of the current thinking by some of our leading practitioners and academics, many of whom are at the forefront of this global revolution. The authors are to be congratulated on the clarity of their writing and the ideas that they postulate for the development of their industry. This book provides not only pointers to the future, but solutions within today's construction practice. It calls on work already being undertaken in UK and international construction and provides a catalogue of techniques and methods for immediate improvement. It is essential reading for all those who want to remain competitive and wish to see major advancement in construction for the benefit of all mankind.

Construction is the world's largest industry. If substantial savings in cost can be achieved, if quality and performance can be markedly improved, and if time for construction can be shortened, then the world will be a better place.

Process improvement coupled with information support is a challenge that cannot be ignored and its benefits go beyond even the profit line of successful construction companies. We believe that those who read this book will find it not only a mine of useful information, but also a stimulus to change.

Professor Peter Brandon
University of Salford

Preface

This edited work brings together the collected thoughts of some of the leading players in universities and construction management and IT companies from around the world. As an edited volume, it inevitably reflects some alternative interpretations on common issues. This is seen as a strength of the collection of chapters. If this field were so simple that there were obvious and straightforward ways forward that everyone agreed with and recognised, there would have been no need for this book to be written.

It is our feeling that there is a pressing need for this book to be read. Its primary target is the informed senior practitioner within construction businesses around the world. It is hoped that some of them will take the time to become acquainted with its messages. Every attempt has been made to present the material in a way that helps this to happen. For instance, a one-minute manager's summary precedes each chapter for the benefit of readers wishing to gain an overview. Ultimately, construction business executives will need to absorb the messages contained within the pages that follow. It is our strong belief that the current business imperative makes this a necessity for business survival.

However, there is a strong cultural barrier to overcome that can be likened to the general on the Victorian battlefield refusing to listen to his strategists offering him a machine gun saying 'I haven't got time for that. I've got a battle to fight!' as he charges into the fray, sword in his hand.

This book has contributions from 27 people from 19 organisations from seven countries in four continents. Seventeen of the 27 might best be described as academics and the other 10 as industrialists. However, such a dichotomy is too simplistic. All of the academics contributing to this volume work in very close partnership with senior industrial collaborators. All of the industrialists have extensive working relationships with academic institutions. We are all hybrids of one shade of grey or another. We all seek to find ways of improving the performance of the construction process through better ideas.

This particular group of contributors sees IT and process

improvement as key ways in which this can happen. We have all found a way of respecting each other's viewpoints, priorities, mechanisms and beliefs. This is the point of scientific and scholarly endeavour. Science and scholarship without relevance and application are pointless. Practice without innovation is terminal.

The major barrier to the effective uptake of the ideas within this volume and the achievement of the mission of this exercise will be the reluctance of industry managers to even listen to its message. The anti-science and anti-academia culture that has prevented the UK exploiting its world-class science over five centuries is renowned. It is still the prevailing culture within UK construction.

Wherever in the world this volume is read and acted upon, it is our sincere hope that the thoughts and suggestions within it prove to be of some value. The publication of this volume is yet another starting point for industry and academia to better understand how we can improve construction processes. This is a continuously moving goal not a destination.

Professor Martin Betts
University of Salford

Acknowledgements

A collected volume of chapters such as this clearly requires thanks to be passed to a very large number of people. This is particularly the case given that the work presented in this book has largely come from collaborative research and enquiry involving teams of academics and industrialists from a number of countries. We are bound to leave some of these important people out, in error, in drawing attention to those that do spring strongly to mind. To those, we apologise and offer our thanks for their contributions nonetheless.

In general, there have been three major groups of people that have influenced the thinking behind these chapters. The first is the group of academics and industrialists in Singapore, and the wider international academic community, who were collaborators in the development in thinking behind much of Part A. These include research assistants working on projects funded by the National University of Singapore including Lim Cher and Lee Tean Jie. Associate Professor Krishan Mathur was also influential in the thinking behind this early work. Members of the international community who influenced this thinking include Matti Hannus, Jarmo Laitinen, Martin Fischer and Yusuke Yamazaki. Industrialists who made inputs to the early thoughts on strategic thinking include Campbell Mackie, John Hemmett and Seah Choo Meng.

The second main group of people that have been instrumental in the development of Parts B and C as a whole are the many industrialists who have contributed so much effort to the work of the Construct IT Centre of Excellence. These were so ably lead in the formation stages by Geoff Topping of Taylor Woodrow and Noel McDonagh. They have been followed with equal enthusiasm and support by Tim Broyd, Matthew Bacon, Martin Jarrett, David Smith, Derek Blundell, Graham Mathews, Adrian Sprague, Chris Powell, George Stevenson and Clive Seddon. This group also includes the excellent support that the authors of specific chapters have received from steering committees of projects that have resulted in some of these chapters. These include: the many contributors to the IT benchmarking working groups; Graham Cunningham-Walker, David Jones, Colin Darch, Bill Bowmar, and David Avery for Chapters 9 and 13; Paul Davis, Jean Michel Doucet, Christophe Gobin and Mats Akerlind for Chapter 12; Marion Carney, George Marsh, Colin Robertson, Claud Brandt and Mark Joynson for Chapter 14; Tony Thorpe, Alan Penn, Eddie Finch, Walid Tizani, Bryan Lawson, Mustafa Alshawi, Tom Maver and Dave Bloomfield for Chapter 16; Richard Baldwin, Matthew Bacon, John Hamspon, John Hinks and Darryl Sheath for Chapter 17; and Tony Bromley for Chapter 18.

The third main group of people we must acknowledge for creating the environment in which this book could emerge are the highly skilful leaders, contributors and facilitators of the research activities at the University of Salford, where much of the work here has come to fruition. These include Peter Brandon, James Powell and Peter Barrett for their leadership and vision, Mustafa Alshawi, Ian Watson, Ming Sun, Grahame Cooper, John Kirkham, Terence Fernando and Yacine Rezgui for their contributions to the overall thinking, and to Sue Buckley, Joanne Nuttall, Michelle Wallwork, Sandra Heyworth, Pamela Allen, Angela Tivey, Heather Vidgen and Jane Tither for creating the conditions that were so conducive to this work being completed so successfully.

Much of the work reported in this book has received financial support from research funders. We are very grateful to the National University of Singapore, the Engineering and Physical Sciences Research Council, the Innovative Manufacturing Initiative, the Department of Environment, Transport and the Regions and the EU Esprit programme for the much needed financial support. Finally we would like to thank Will Swan, Stuart Birchall, Nick Bakis and Simon Osbaldiston for keeping the technology working for us, Will in particular for his assistance with the production of some of the artwork, and Joanne for creating the time to allow it to be finished.

Contributors

The Editor

Martin Betts Professor Betts is a Professor of Construction IT and the Head of the five-star rated School of Construction and Property Management at the University of Salford. He was, until recently, the Director of the Construct IT Centre of Excellence. This is a UK government-backed national network of some 50 companies and institutions and ten universities and research institutes who are coordinating their research and innovation in construction IT for the competitive benefit of the construction industry. The centre is conducting benchmarking of IT support to business processes, developing methodologies for measuring benefits in IT investment, showcasing emerging technologies, communicating IT innovation, and defining a long-term research work plan. Professor Betts was also the initial Programme Director of the EPSRC-funded Integrated Graduate Development Scheme at the University of Salford.

Professor Betts has extensive international experience having lived and worked for five years in South East Asia. He is also the coordinator of the Conseil International du Bâtiment (CIB) Working Commission 78 concerned with IT in construction. This group of more than 200 experts in construction IT around the world conducts ongoing international collaborative projects, publishes an electronic journal and holds major annual workshops. Professor Betts has major research interests in the strategic management of IT at the corporate and national levels and has published extensively in major journals and conferences throughout the world. He is a visiting professor at the Royal Melbourne Institute of Technology in Australia.

The Authors

Ghassan Aouad – University of Salford Dr Aouad is the Associate Head of School for Research of the School of Construction and Property Management at the University of Salford. He is a Reader in Construction IT and Management at the university. His background is in intelligent integration of information in construction management. He is conducting major research in the field of integrated project information and project process improvement. He is lead researcher in the Open Systems for Construction (OSCON) project at Salford and a lead researcher on the Generic Design and Construction Process Protocol work. He is the author of many major journal papers in the field.

Brian Atkin – Royal Institute of Technology, Sweden Professor Atkin holds the chair in Construction Management and Economics at KTH in Stockholm and is a Director of Atkin Research and Development Limited. Atkin R&D made a substantial contribution to the many benchmarking studies reported in this book. He was formerly a professor at the University of Reading. He leads the major Intelligent Manufacturing Systems global research initiative FutureHome. Professor Atkin's interests are wide-ranging in the fields of construction management, construction IT, robotics, value management, facilities management and intelligent buildings.

Andrew Baldwin – Loughborough University Dr Baldwin is the Head of Department of Civil and Building Engineering at Loughborough. He has extensive industry experience in the design, development and implementation of computer-based information systems for construction management. His current research interests focus upon work process changes in the construction industry and their impact upon construction organisations.

Derek Blundell – Construct IT Centre of Excellence Mr Blundell is the manager of the Construct IT Centre of Excellence and an independent consultant to a number of major UK contractors on the subject of IT management and strategy. He has many years of practical experience within Trafalgar House Construction as a Planning Engineer, Construction Director and Business Systems Director. In his capacity as Business Systems Director he has experience of developing and implementing IT strategies for 18 business units within a major international construction organisation.

Andy Clark – UNISYS Mr Clark is a former director of the Construct IT Centre of Excellence. He currently works as a consultant within the UNISYS Construction Solutions Business and is responsible within the company for helping ensure that a major IT supplier is able to present IT products in line with the IT strategies of a number of major international construction organisations.

Rachel Cooper - University of Salford Professor Cooper holds the chair in Design Management at the University of Salford. She leads a research group in Design and Manufacture. Professor Cooper is chair of the European Academy of Design and the editor of the *Design Journal*. She has taken part in an 18-country study of new product development in high technology industries. She has also undertaken a US/Japanese/UK comparison of the role of suppliers in design in the automotive industry. Since 1995 she has led funded projects resulting in a generic design and construction process and guidelines for partnering in construction. Professor Cooper has published four books and over 100 academic papers. Her latest book is in the field of marketing and design management. She is currently the lead investigator developing generic design and construction sub-process in conjunction with nine industrial partners including Alfred McAlpine, BAA and BT.

Paul Davies - Building Design Partnership Mr Davies is the IT Manager of BDP, the leading multi-disciplinary design practice in the UK. His responsibilities embrace developing a strategic vision of how IT can support the business strategies of his business. He is also responsible for a team of people who implement this strategy and support IT implementation and application throughout the BDP business. In 1998, Mr Davies was a nominee for the CIOB IT Manager of the Year Award in the UK.

Anne Marie Dubois - Centre Scientifique et Technique du Bâtiment, France Ms Dubois is a Senior Researcher at the leading French research institute CSTB. She has extensive experience of research in the fields of product and process modelling and IT support in construction. Her current research interests are in technology transfer mechanisms and the applicability of research results to construction practices.

John Gravett - Atkin R&D Mr Gravett is a research consultant to Atkin R&D. He is undertaking industrial research in the areas of IT support in construction, design and build and supply chain management. His previous experience is as a Senior Quantity Surveyor within Gardiner and Theobald in London. Within the practice he had a management responsibility for research projects.

Antonio Grilo – FORDESI Dr Grilo has recently completed a PhD at the University of Salford in the field of factors influencing the development of electronic trading in construction. He previously completed the MSc in IT for Property and Construction at the university. He is now a senior manager at FORDESI, a growing firm of management and information systems consultants in Lisbon, Portugal.

Karen Lee Hansen – VANIR Construction Management, Inc. Dr Hansen is a Senior Project Manager with a leading California-based program, design and construction management firm. She completed her doctorate at Stanford University in the Construction Engineering and Management Program in strategic decision making. She has extensive industrial experience as a manager within The Boeing Company and in general construction. She has conducted research at the Strathclyde, Sussex and Brighton Universities in the UK. Her research interests include complex product systems, process improvement, technology transfer and design methodologies.

Alan Hutchinson – AMEC Construction Mr Hutchinson is Quality Manager within the AMEC group of companies. He chairs the steering group of the Standardised Performance Improvement for Construction Enterprises (SPICE) project. This is a major UK government funded initiative to apply principles of the Capability Maturity Model to performance improvement within construction supply chains.

Adina Jägbeck – Royal Institute of Technology, Sweden Dr Jägbeck has recently completed a doctorate in product modelling in construction at KTH in Stockholm. She has previous industrial experience as an architect working at VBB, today part of SWECO. Her current work involves working with the IT Bygg Construction IT Implementation Centre in Stockholm.

Martin Jarrett – Bucknall Group Mr Jarrett is the Group Technical Director of the Bucknall group of companies. He is leading a major business process change activity within the group and is responsible for the group's IT strategies to help implement this process change. He is a founding member of the management board of Construct IT Centre of Excellence and has led several of its major projects and assignments. He was formerly managing director of a software house and successfully implemented financial management systems within a number of major UK consulting organisations.

Mike Kaglioglou – University of Salford Mr Kaglioglou is a Research Fellow within the Technology, Information, Management and Economics Research Institute at the University of Salford. His background is in manufacturing management and engineering management. He plays a major role in the ongoing work in developing detailed generic process protocols for the construction industry in conjunction with leading industry innovators. He is also pursuing research interests in the field of new product development.

Rachael Luck – Reading University Ms Luck is a Research Fellow within the Department of Construction Management and Engineering at the University of Reading. She has extensive research experience in the fields of project integration, construction futures and design management. She is an architect and council member of the Design Research Society.

Denny McGeorge – Newcastle, Australia Professor McGeorge is Acting Dean of the Faculty of Architecture and Building at the University of Newcastle, New South Wales, Australia. He has extensive experience of research in the fields of building performance, construction IT, value management and risk management. He is a leading authority in the field of problem-based learning in construction. He has held a number of visiting professorships which recently include Hong Kong Polytechnic University and Glasgow Caledonian University. He was a member of the Royal Commission into the NSW Building Industry. He is a co-author of a recent book on new management philosophies in construction.

Marcela Miozzo – UMIST Dr Miozzo is a lecturer in Technology Management at the Federal School of Management in Manchester. She has completed a PhD in Economics for the University of Massachusetts at Amherst, USA, for which she obtained a Fulbright Grant. Her main interests are in the fields of industrial regulation and policy and the economics of innovation. Current work involves the study of implications of recent technological and organisational transformations in the international automobile industry in South America. Following her research work with the Construct IT Centre of Excellence, she is currently studying the implications of recent changes in project management and technology for the UK construction industry, under a European Regional Development Fund award.

George Ofori – National University of Singapore Dr Ofori is an Associate Professor within the School of Building and Real Estate at the National University of Singapore. He is a leading international scholar in the field of construction industry development. He has conducted major studies for the International Labour Office, the United Nations Centre for Human Settlements, and the Construction Industry Development Board in Singapore. He is the author of a number of major books in the field of construction industry development and the economics of the construction industry. He coordinates the new CIB Task Group 29 on Construction in Developing Countries. His current research interests include sustainable construction, construction industry development and construction in developing countries.

Martin Ridgway – MSW Technology Mr Ridgway has extensive experience of working within IT and management consultancy within construction and other sectors. His previous experience includes major periods and assignments with Rider Hunt, Henry Riley Group, Tarmac Construction and Coopers and Lybrand. After completing an MSc in IT for Property and Construction at the University of Salford, he has taken a position as Implementation Manager with MSW, a leading software house.

Marjan Sarshar – University of Salford Dr Sarshar is Director of the Construct IT Centre of Excellence based at the University of Salford. Her background is in the IT industry. She previously worked as a researcher at Salford on projects related to integrated information systems for construction. She now leads major projects in the field of IT and IS strategies and standardised process improvement in construction enterprises (SPICE). In this work she has major links with senior representatives of UK construction companies and with leading international research groups.

Martin Sexton – University of Salford Martin Sexton lectures in construction management in the School of Construction and Property at the University of Salford. His recent research projects include work in facilities management, quality management in higher education, process management, innovation management and sustainable development.

Mathew Shafaghi – Farnborough College Dr Shafaghi holds a doctorate in IT systems development and implementation in manufacturing from Sheffield Hallam University. He has worked for a number of years as a Research Fellow at the Construct IT Centre of Excellence. In this time he developed close working relationships with a number of major UK and international construction companies. Dr Shafaghi is currently a lecturer in the schools of Computing and Information Systems and of Management at Farnborough College of Technology. His academic post is complemented by his relationships with SMEs and advising them on various aspects of IT management.

David Smith – UNISYS Mr Smith has been working for 20 years in developing improved application of IT within construction. Part of this time was working with Wimpey who at the time were a major international building contractor. His more recent experience has been with the major global systems integration company Unisys. Until recently, Mr Smith led their global construction solutions business. He is a founder member of the Construct IT Centre of Excellence management board and leads its benchmarking activities. His new position within Unisys sees him working in client relations in Unisys' work in the insurance sector.

Tony Thorpe – Loughborough University Professor Thorpe holds a chair in Construction Management in the Department of Civil and Building Engineering at Loughborough University. He has extensive experience in basic and applied research in applications of advanced technologies within construction companies. He is currently engaged on major research projects with collaborators from leading international construction firms, in the fields of construction communications, construction process improvement, design management, briefing and collaborative working.

Stephen Walker – Tilbury Douglas Construction Mr Walker is employed as an IT strategy analyst within Tilbury Douglas Construction. His role within the company is to develop long-term strategic vision for IT within the company, inform investment decisions and drive business improvements. Responsibilities include ensuring investment proposals are produced that meet business needs and the management of subsequent implementations. Mr Walker was previously engaged as a researcher within the Construct IT Centre of Excellence. In this role he liaised closely with a number of major construction companies in benchmarking their IT support to a full range of construction processes. Mr Walker's background is in construction management within Tilbury Douglas.

Strategic Management and Business Process Analysis

This first part of the book sets the context for why the new enabling technologies of IT and communications are of such importance to the modern construction business. It introduces the concepts of business strategy and strategic management and relates them to the construction sector. It also illustrates how the new process paradigm is so important to modern strategic business thinking.

It shows the importance of IT as an enabler to such developments and introduces techniques of strategy development, IS planning, IT planning and IT implementation. It concludes by outlining, in general terms, why and how such thinking must be taken further in construction.

This first part of the book contains a combination of theoretical concepts and ideas, generic descriptions of techniques, and specific practical guidance. The chapters in this part, in common with other parts of the book, are illustrated by a series of unattributed, real-life

quotations from senior managers from construction businesses around the world.

Some real-life case studies are also included in an attempt to bring to life the concepts introduced in the chapters. For the benefit of readers wishing to gain an overview of each chapter, a one-minute manager's summary precedes each chapter in this and all other parts of the book.

1 The importance of new approaches to management

Martin Betts, Andy Clark and George Ofori

"We will have increasingly to apply sophisticated technology in doing our work and to understand what our clients need. For those who adapt to meet the changes the future is bright."
– *Executive vice president of US-based global engineering constructor*

The business world is fast changing. This causes all companies to rethink what markets they should be in and how to develop products and services to serve these markets. Construction is no different. Because of the relatively slow change in construction in the last 30 years we can expect it to change even more dramatically than other sectors in the next 30 years. Information technology and redesigned business processes will be central to this change.

- Products and services of industries like retailing, banking, insurance and leisure are unrecognisable from ten years ago.
- New priorities affecting construction are strategic orientation, quality, deregulation, environmental concerns, external competition, client demands and the rate of technology change.
- Approaches to management have been fundamentally changed by thinking about processes rather than functions.

"The culture of industry needs to be changed. The fragmented nature of industry may be the major barrier to change."
– *Managing director of large UK contractor*

1.1 Introduction

The world was once a fairly straightforward place where supermarkets sold food, petrol stations sold petrol, and the bank was the old building on the high street where you went to manage your finances.

Now, supermarkets sell petrol and banking services, petrol stations have become supermarkets and the old building in the high street is now more than likely part of a chain of restaurants and bars. What is more, this establishment is unlikely to be an escape into traditional values and methods from the way the world has changed. The bar probably has sophisticated information management systems integrating cash management, financial accounting, marketing and distribution such that menus, stock and promotions can be rapidly changed to reflect a change in taste, fashion or season. Such information management systems have become the key to the way the modern food and drink outlet implements its business strategy and thrives and competes in the modern business environment.

Similar information management systems have radically altered the nature of the banking sector. From a business that relied on personal service, a physical presence, a feeling of tradition and opening hours that suited the staff rather than the customers, the modern bank has been transformed. It is now more likely to be a 24-hour, telephone-based operation from a warehouse in the industrial suburbs of a major city. It is probably staffed by former nurses, firemen and others who are used to working shifts around the clock and have an ability to communicate their genuine care for people through the telephone.

This change in the banking sector was first triggered by the emergence of the automatic telling machine (ATM) as an electronic extension to the physical operations of the branch. The story of how the First Bank of America emerged from holding a small market share to becoming a company that was a major player in global banking on the back of its information system based delivery of its business strategy shows how IT and IS have transformed the banking sector.

The nature of business in all sectors has undergone accelerating change over recent years. The nature of relationships between participants in the supply chain of business processes has moved from adversarial to cooperative. The attitudes towards customers has been transformed towards a situation where developing a supportive input to a customer's processes is seen as imperative to the modern business organisation. Within organisations, management thinking has evolved from considering functions and specialisms to thinking of business processes. Contextual drivers to modern business have been radically influenced by the emergence of environmental concerns and the quality movement.

These developments in business have been accelerated by the emergence of the information revolution. The emergence of affordable computer processing power initially unleashed an explosion of new business support systems addressing the internal processes of business organisations. Such a phenomenon is now being made almost insignificant by the newly emerging developments in availability of high volume and affordable bandwidth for telecommunications and the explosion of the Internet.

Some of the implications of these developments are the increasing outsourcing of some of the activities of information management within organisations, an increasing pattern of teleworking and hot-desking within employment patterns, and the effective management of buildings and other facilities rising on the corporate agenda.

1.2 The new business priorities

The need to apply a strategic perspective to business operations has been recognised for over three decades. More recently, in economics and business management, frameworks and priorities have shifted to a greater extent from the short-term and tactical to the long-term and strategic. This shift has been at both corporate and national levels, in response to challenges to business environments caused by the increase in global competition in almost every industry. These changes have caused all stakeholders in economic activities to reappraise their attitudes, not only to products but also to the long-term relationships between buyers and suppliers.

Back in 1979, M.E. Porter described how changes in the business environment were leading companies to reconsider their views on 'traditional' economic theories of competition. The theory of perfect competition assumes:

- free access to new participants in an industry; and
- the availability of perfect market information to all within a particular sector.

The extent to which these principles apply in practice affects the profit potential. Porter suggested that a time of rapid business change enables managers to examine how much they can advantageously influence the level of competition within their industry and therefore gain abnormal profits. This is contrary to the usual perception that rapid change represents only a threat.

Other factors are also important. According to Davenport and Short (1990), the quality movement, in manufacturing and other sectors, led organisations to start analysing their total business process instead of their individual tasks or units. They suggest that major redesign of whole business processes is occurring as part of a new industrial engineering climate. We will return to this theme in the final part of this chapter. In 1984, Ives and Learmonth highlighted deregulation as a major cause of the change in emphasis in business. They also identified global competition and declining cost of new technologies as important stimuli. In addition, there are clear changes in access to industry and particularly access to markets in construction, as we will show later.

In conditions of rapid change, companies and business managers can no longer rely on a strict return-on-investment (ROI) evaluation of business ventures of a tactical nature. They must explore value-added concepts of a strategic nature. The use of ROI as a criterion for business decision-making is inappropriate in view of the changes that have occurred in modern business environments. The implementation of the first ATM machines by banks would not have been justifiable on ROI criteria alone.

An ROI focus by senior management may turn attention toward narrow, well-defined targets as opposed to broader strategic opportunities that are harder to analyse. Schwartz (1991) argues along these lines in his proposals for taking the long view of economic situations. His work in economic planning for Royal Dutch/Shell is based on long-term strategic planning. For the construction industry, we will make a clear case for IT to be viewed as part of a long-term strategy - going beyond a mere consideration of its short-term financial returns.

1.3 The dynamism of the construction sector

In the context of all the dramatic changes we have described above, construction appears as an historical beacon of traditional attitudes, methods and approaches. This cannot continue to be the case. Developments in patterns of working, including teleworking and hot-desking, are already reducing the level of demand for buildings in the long term. The greater emergence of virtual retailing and entertainment will add to this. The threat of substitutes to the products and services of the construction sector is increasing. The impact of developments in environmental awareness is also increasing.

Construction activity is subject to:

- influences resulting from the pace of technological change in other sectors of the economy;
- increasingly stringent regulations; and
- changing client desires as a result of variations in tastes, aspirations and purchasing power.

Yet the most pressing need for change confronting construction at present is the fact that markets in major developed economies have become saturated with similar companies offering over-capacity of undifferentiated products and services to a largely dissatisfied client base. Whilst the level of international competition is yet to reach the fully globalised nature of many other industries and markets, it is increasing and there is a trend for more foreign ownership of domestic construction firms in different countries.

Partly as a response to these external influences, competition within the construction industry is increasingly intense and sophisticated. Firms adopt practices and procedures which help them to survive, but also, the 'normal' competition among enterprises in particular segments of the industry is being overlaid with competition in other forms. Moreover, there is now external competition – manufacturers of construction materials and components are integrating vertically by offering construction services, in some cases, as front-end loss leaders.

In this increasingly dynamic situation, it is clear that construction enterprises will have to be vigilant and forward-looking to survive, let alone to do well. Tactical considerations will need to be replaced by, or at least put in the context of, strategic ones. Addressing a conference of the International Federation of Consulting Engineers in 1989, Cordel Hull, Executive Vice President of Bechtel, observed:

"We will have increasingly to apply sophisticated technology in doing our work and to understand what our client industries need... To develop engineering packages and give clients a total service, that is the key.... Be flexible, stay at the leading edge. For those who adapt to meet the changes the future is bright."

Up till now not many construction enterprises appear to be addressing the issues. In 1991, *The Economist* examined some of the build-operate-and-transfer projects of Hong Kong based Hopewell Holdings (a $560 million power station in China's Guangzhou Province; a six-lane toll-road between Hong Kong and Guangzhou that opened in 1993; plans for a 500km highway in the interior of China and a 60km network of elevated and light roads in Bangkok, Thailand). In the UK, the Private Finance Initiative (PFI) is emerging as the main way in which UK schools, hospitals, roads and other facilities are being procured. The ambition and financial ingenuity underlying these projects are great. Not all companies are following this trend because it is risky and nerve-racking work which often requires several years of lobbying, and the right political connections and the willingness (and ability) to use them. More conventional construction firms try to avoid involvement in politics, but with estimates showing that the countries in East Asia will have spent over US$1 trillion on infrastructure works between 1991 and 2000, they have been tempted to change their minds. The more recent slowdown in that part of the world has only made players in those markets seek to be even more competitive. In an industry that forms 10% of global GDP, new opportunities emerge everywhere.

The nature of construction activity, its structure and its operating environment are fluid and dynamic. This dynamism is growing at an increasing pace, offering proportionately greater strategic opportunities, while posing significant threats. Clients are becoming more aware of the nature of construction (and of their rights) and tending to be more discerning and demanding. With increasing statutory control, greater client and user knowledge and, especially, concern with the environment, professional duties and liabilities are being defined more strictly.

There is also increasing deregulation and further privatisation relating to development and building control. In many countries, mandatory professional fee scales and statutory protection of professions are already things of the past: elsewhere, they are under threat. The challenges and opportunities offered by this situation are well illustrated by changes to the statutes relating to construction in Singapore. The Building Control Act 1989 imposed greater responsibilities on architects and engineers (in the design and supervision of building works), and contractors. In response, many companies have been formed which undertake newly required statutory periodic structural inspections (not more than five years for commercial buildings and not more than ten years for residential buildings) and, when necessary, the repair of buildings used by the general public. In addition, the statutory changes have led to the Professional Engineers and Architects Act 1991 in Singapore which allows the formation of limited liability companies by construction professionals and multi-disciplinary enterprises.

The environment of the construction industry is being increasingly influenced by economic, technological and social factors. For example, economic forces, within and outside a specific country, include increasing privatisation and globalisation. This is exemplified by the rising importance of regional economic co-operation – reduction of trade barriers and harmonisation of regulations. In general, these lead to greater competition from firms from outside the country concerned. Another major influence comes from the progress towards the information age. A structural re-organisation of the industry appears likely. As the way in which we execute construction projects undergoes radical changes, those who position themselves strategically can maximise their benefits.

We have examined construction enterprises in various countries and reviewed their strategic responses to the dynamism of their operating environments. We found that some companies have diversified their activities to become integrated architecture, engineering, construction (AEC) firms. This follows an involvement in design-and-build and turnkey projects or through mergers, acquisition and strategic alliances both within and across national borders. Others have ventured outside construction. Business development has become an important part of the operations of the construction enterprise. Competition on the basis of far more than price is increasingly important. Build-operate-and-transfer arrangements, such as UK PFI schemes, are examples of the result of this phenomenon. Contractors' tenders now commonly incorporate financial packages; the contractor is joint venturing not only with other contractors but also with the developer; and more recently, some contracting firms have offered such services as the identification and securing of major tenants for proposed developments.

It is clear that many changes in the environment of the construction industry have implications for specific professions and trades which go beyond what the individual practitioner or enterprise can effectively address. Perhaps more than other sectors of the economy, strategic management is of concern to organisations in the construction industry other than enterprises.

The professions and the trade associations within construction industries must also respond to strategic opportunities and position themselves advantageously by examining their internal activities and external relationships with others. Times of great change offer opportunities for competitive advantages, rapid growth and high profit. One needs only to look at the IT industry to see that. Partly as a response to these external influences within the construction industry, professional groups in many countries are working to maintain or expand the role of their members on projects, increasing their market share.

Construction practitioners and enterprises are now under threat from individuals and organisations in other fields of activity that offer construction-related services. These fields include those:

- that construction practitioners have ignored;
- for which they lack the relevant expertise; and
- which, while being part of their traditional activities, they do not perform satisfactorily.

For example, some investment analysts and accountants are offering consultancy services on how building owners can optimise the revenue-earning potential of their property, and how they can most beneficially (synergistically) integrate their property holdings into their corporate assets and operations.

Reflecting the dynamic times within which we operate, some of the current thinking in IT is moving away from solely technological concerns to a greater emphasis on how technology can be managed and applied. The primary purpose of this book is to assess whether these changes apply in construction and whether the way IT is being used in construction demonstrates this.

1.4 The role of information in the construction industry

Information is a key resource within any business activity. Researchers, software developers, and practitioners are now applying information technology to automate different parts of the construction process. The use of IT in construction is becoming increasingly sophisticated with virtual reality, knowledge-based systems, object-oriented approaches and neural networks among the latest technological advances. The knowledge-based applications embrace design, engineering, management and economics. The purpose of these applications, and the thought given to the business and management implications of technology, are less advanced. There appears to be a reluctance to recognise that IT can exploit strategic opportunities for new ways of doing things, not just automating current manual processes. There is also an apparent reluctance to view opportunities for IT use across the whole construction sector and not within individual parts of the process. A major aim of this book is to analyse how we can be more strategic in our use of IT across the sector and how taking a process-based view of our activities allows us to do this.

1.5 The need for redesigned business processes

A glance through the mainstream management press over the last five years will reveal myriad articles describing, discussing and subscribing to process-oriented theories. These theories aim to reorganise the way companies operate to facilitate radical improvements to key measures of performance. These process-based theories have been called **business process redesign**, **process innovation** and **business process re-engineering**, among others. To avoid unnecessary jargon, which often only differentiates theories by nomenclature, we will use the term business process redesign (BPR). The phrase 'business process' focuses the subject of the analysis and the word 'redesign' focuses the aim of studying the process.

The central argument for the focus on business processes is the evolution of organisational structure and management philosophy since the pioneering work of Adam Smith and Frederick Taylor. Smith advocated task decomposition to the smallest constituent level, using the manufacture of pins as an example. He replaced a group of ten generalist pin-makers by specialists, each carrying out one or two of the basic 18 tasks of pin manufacture. This change in the organisation of labour allowed production to increase by a significant factor as each worker's dexterity improved by focusing on a few tasks instead of all 18. Combined with the labour-saving machinery of the Industrial Revolution, these automated specialist tasks became a blueprint for mass consumerism and the work of Frederick Taylor.

Taylor developed the principles of scientific management in response to his personal dislike of waste in the workplace. Scientific management theory examines tasks with the doctrine that workers are unable to find a more efficient method of executing a task due to their familiarity with the skills that could have taken years to perfect. Taylor proved that there were more efficient methods of executing tasks by closely monitoring the performance of workers. He would follow workers with a stopwatch and analysed bodily movement to develop the most efficient method of performing the task in a manner that optimised speed and minimised effort. From this point he developed control and measurement methods for managers to ensure that work was consistently efficient, while providing data to develop further improvements to productivity.

The obsession with reducing the toil of the labouring body led Gilbreth, an associate of Taylor, to develop an adjustable scaffold for piling bricks. This development, which stopped the bricklayer from continually bending over to reach the next brick, raised productivity at the time from 1000 to 2700 bricks laid per day.

The work of Smith and Taylor has been important in shaping a task focus within modern management. Managers have developed a huge knowledge base of the most efficient method of performing a task. While this is admirable, it may adversely affect our understanding of the interaction of tasks that form the process. In centring improvement around the part, we lose sight of the whole. Fine-tuning the work of each employee may therefore sub-optimise the operation of the overall process to which the task contributes. As individual tasks are performed to greater levels of efficiency, companies still encounter workflow bottlenecks not within the task, but between the tasks. Task-centred skills mean that a piece of paper finds its way into an out-tray quicker. Unfortunately it also implies that a speedy passage to the next in-tray is by no means certain. The need for redesigned business process calls for knowledge of the interdependencies that produce a product for a customer, whether internal or external. It is these interdependencies that are part of our current productivity crisis and the reason that current management thinking has turned its attention to processes.

The processes that occupy the majority of companies are the evolutionary consequence of the fragmentation of work. Company structures that are organised to focus on the task are functionally led by specialist departments and numerous hierarchical levels. This hierarchy provides a framework to deal with any eventuality and control mechanisms to halt any abuse. Detailed rules that dictate procedure and behaviour constrain employees to work in a determined manner, leaving little scope for free thinking and innovation. Although relevant to the era of mass consumerism, low competition and a narrow product differentiation, this structure needs fundamental change to arm companies with the weapons to succeed in the next millennium. Now the environment is one of global competition, cyclical economies and rapid advances in technological development. The processes that currently dictate the passage of work to produce a product or service for a customer are incoherent and in need of redesign for the tasks to integrate to optimise process performance.

The shift of focus from task to process implies that companies concentrate their structure around the output to the customer. A process viewpoint examines the input(s) to the process and asks whether value is being added at each step of transformation to produce an output. Traditional functional wisdom streamlines organisations by cutting the operating costs of departments while process thinking asks the question 'How can I add more value for less effort?' Streamlining processes to deliver a better product, over a shorter timescale, for a competitive price naturally reduces costs. Waste activities are discarded when put up against the test of whether or not they add value.

Johansson *et al.* (1993) define a process as 'a set of linked activities that take an input and transform it into an output'. All companies have processes, but many will have developed in an *ad hoc* manner within the functional structure. For many companies the agenda is not in business process redesign, but in business process design. It could be argued that most processes were never designed to be redesigned. BPR as a concept and methodology is examined in more detail later in this book.

1.6 Summary

The shift in emphasis in management in organisations from tactical to strategic is clear. There is also a shift from internal concerns with narrow performance criteria such as ROI (return on investment) to external concerns with concepts such as value and competition in global markets. Enterprises, groups of firms and national agencies in many sectors have realised that it is insufficient and often dangerous merely to react to events and are endeavouring to influence the future, at least as it relates to their operations.

The construction context is becoming increasingly dynamic. All construction organisations need to think differently about how they conduct their business. Technology, regulations and market changes are causing new strategic responses. Competition is coming from more unexpected sources. Information is becoming a key competitive resource.

2 Strategic management – definitions and techniques

Martin Betts, Andy Clark and George Ofori

"The construction sector revolves around exact information."
– *Head of construction, large Belgian contractor*

There are a range of techniques that can be used to aid in the planning phase of strategic management. These allow a company to analyse the industry and markets they are in, distinguish which competitive strategy they are following, understand why groups of companies in certain sectors and countries do better than others, benchmark themselves against other companies, identify and exploit core competencies, develop strategic visions and redesign their processes. All these techniques require companies to adopt modern business process thinking. They can be used to help a company in construction to identify new ways in which they can approach construction markets.

- The strategic health of different industries and markets is inherent from their underlying features. This can be analysed and influenced.
- Companies can follow different strategies to compete in different markets depending on whether they seek to differentiate or be the low-cost provider in broad or narrow market segments.
- Companies can act strategically by developing core competencies, developing strategic vision and redesigning their business processes.
- The fundamental underpinning of modern business strategy is business process thinking.

"If the right information does not get to the right person, work may stop or go wrong. With project fixed costs of FF10K per hour, this can make the whole project, or the whole company collapse."
– *Technical director, large French contractor*

2.1 Introduction

In Chapter 1, we saw the importance of strategic as opposed to tactical management. But what is strategic management? We begin this chapter by considering some specific techniques and models for strategic management, notably those proposed by Michael Porter. We then introduce the principles of corporate benchmarking and look at some other alternative approaches to strategic management. Finally, we examine in some detail the principles of business process redesign.

2.2 Strategic management techniques

In line with the changes in business thinking has been the emergence of many new tools and techniques for business planning. Many of these techniques are aimed at specific management functions such as marketing, human resources, and technology. Others are of a more general nature and are intended for overall corporate strategic management activities. One of the most powerful concepts to emerge was of a sustainable competitive advantage (SCA). Early work by Michael Porter explored the issues of strategic management by enterprises for competitive positioning. The concept of developing an advantageous competitive position that could be maintained by the enterprise for long enough for it to make substantial gains was referred to as an SCA.

Porter also developed a number of tools and techniques that could be used in identifying an SCA. These have been built on by others who describe an analytical framework for strategic business management that entails:

- defining an organisation's mission;
- identifying its goals and objectives; and
- deriving strategies for alternative scenarios.

The framework, which involves the use of the 'five forces' model, 'value chains' and the 'three generic strategies', together with techniques suggested by other authors are applied to construction in this chapter. The range of techniques also shows how IT can be used to:

- gain competitive advantage;
- allow new businesses to emerge; and
- change industry structure.

Some have disputed whether large, modern enterprises can realistically attain SCAs. Their more recent approach suggests that a company should concentrate on progressively improving its core competencies (its key abilities and skills). They argue that the goal of an SCA is now outdated and unrealistic because of the increasing dynamism of the business world. Instead, they suggest that the need is for the enterprise to continuously regenerate its strategic plan to produce a series of short-term but unsustainable competitive advantages. This may well be the case in the more dynamic and advanced economic sectors. We should remember that in the construction sector, application of strategic management techniques lags behind that in other sectors. It seems possible that the concept of an SCA may still apply.

2.2.1 The five forces model

The five forces model originated from work by Michael Porter in which he showed that corporate strategy concerns positioning an enterprise in relation to these forces, influencing their balance within an industry and exploiting industry change and its effects on the forces. The five interrelated forces are:

(1) the threat of new entrants;
(2) the power of suppliers;
(3) the power of buyers;
(4) the threat of substitute products and services; and
(5) jockeying for position amongst industry members.

The interrelationship of these forces is shown in Figure 2.1.

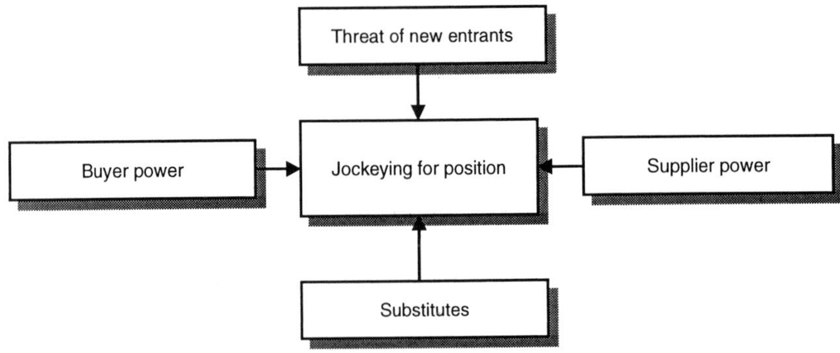

Figure 2.1 The five forces model

The model was first developed by Porter as a tool for industry analysis. It can be used to compare relative potential for superior business performance between different industries and industry segments. Porter identified new product development in the pharmaceutical industry as a 'five-star' performer because of the favourable position of all five forces. He also showed how the airline business moved from an attractive industry from a five-forces analysis to an unattractive one because of deregulation in the 1980s and how airlines used frequent flyer programmes, the hub and spoke concept, and IT-based ticketing systems to change the forces back to being a strategically attractive industry.

The technique can readily be applied to strategic management in construction. It enables a construction enterprise to analyse product and location segments of the construction market and to identify areas where superior business performance is more likely because of the more favourable position of the five forces. It is therefore of fundamental strategic importance in helping an organisation decide whether it is in the right business.

Threat of new entrants

Porter later advocated use of the model by an enterprise wishing to change and influence the current balance of the forces within an industry segment. He observes that the threat of new entrants within an industry can be influenced by factors including:

- economies of scale;
- product differentiation;
- large capital requirements; and
- cost disadvantages associated with learning curves and experience curves.

The threat of new entrants is particularly potent within the construction industry: construction is noted for the ease with which new firms can enter its different parts with minimal investment.

But consider the increasingly large investment by major construction firms in IT and engineering technology. It could be argued that this is providing a barrier to new entrants in the larger project and high-technology end of the construction sector in the same way that pharmaceutical research and development infrastructure is a new entrant barrier. Capacity and experience for PFI (public finance initiative) projects may also emerge as a barrier to entry to that market.

Power of suppliers

The competitive advantage to be gained from powerful supply relationships is greatest:

- when few enterprises dominate supply;
- where there is no competing product; and
- when the supplier holds a threat of forward integration over the buyer.

In the construction industry, within each country, region or other relevant geographical area, there are often many different suppliers of many products and services. Major contractors at present seem to merely offer undifferentiated capacity of an identical nature to each other. Yet the scope for competitive advantage in some of these situations could be exploited.

Power of buyers

Conversely, a buyer has greatest competitive advantage when:

- it is a large volume buyer;
- it purchases undifferentiated products that are price sensitive but not quality sensitive to its processes; and
- when it holds a threat of backward integration over its supplier.

There are many opportunities for large construction enterprises to exploit this opportunity with materials suppliers and specialist subcontractors. The emergence of both design-and-build and turnkey constructors are examples of the extent to which this competitive advantage may be exploited. In Sweden, we shall see later how Skanska have exploited this buyer power opportunity and used IT to support this business strategy.

Threat of substitutes

The potential for substitutes gives scope for competitive advantage within an industry. This applies particularly when they are subject to trends improving their price–performance trade-off and when they are produced by industries earning high profits. An example cited by Michael Porter (1980) relates directly to construction. He refers to the household thermal insulation market as one where there was the possibility of abnormal profits in the fibreglass insulation industry in 1978. This opportunity was quickly removed by the emergence of substitute products including cellulose, rockwool and styrofoam. A competitive advantage quickly proved to be unsustainable.

The threat of substitutes to construction is being substantially triggered by developments in IT. The need for road and railway infrastructure is being influenced by the emergence of communications technologies and their enabling of hot-desking and teleworking influences demand for buildings as a substitute.

Jockeying for position

Jockeying for position by existing industry members is a fifth force described. It gives greatest scope for competitive advantage when the following conditions apply in mature, saturated markets:

- many competitors of equal size;
- slow industry growth;
- undifferentiated products or services;
- high fixed costs in enterprises; and
- high exit barriers from the industry.

Parts of the construction industry, especially at the lower end of the technological scale, do have large numbers of similarly sized competitors with undifferentiated products and services. Yet fixed costs tend to be somewhat low. This, together with low exit barriers, causes a high entry and attrition rate amongst construction enterprises. Despite this, the jockeying for position of existing members is a strong competitive force within construction. It is clear that each of these five forces is directly relevant to construction. Yet, there is little evidence of construction enterprises systematically examining these five forces and their relevance to the strategic management of their firms or to their targeted deployment of advanced IT.

2.2.2 Value chains

We have seen that business planning developments have led managers to stop looking at discrete activities within firms but rather to consider firms as a whole. Yet Michael Porter argues that to identify potential for competitive advantage we must look at individual parts of the whole firm within its supply chain, preferably using the value chain. This is a structured way of analysing the constituents of a business in the categories shown in Figure 2.2.

Figure 2.2 The value chain

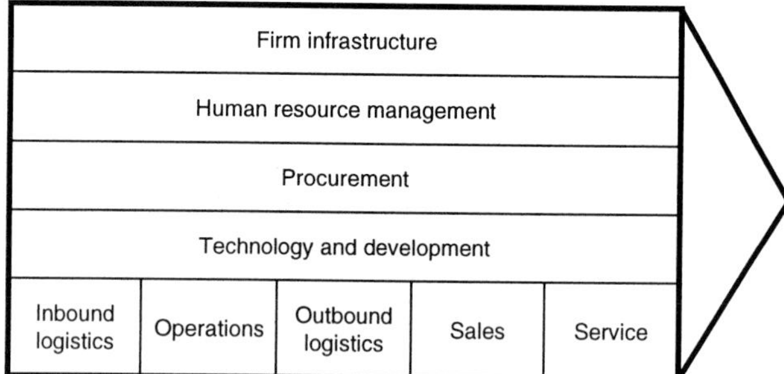

Differences between value chains are a key source of competitive advantage between competitors, and the importance of technology to value chain analysis should be stressed.

The value chain divides an enterprise's activities into technologically and economically distinct value providers. Value is what a company creates, measured by the amount buyers are willing to pay for a product or service: the difference between value and cost determines a company's profitability. Porter concludes that competitive advantage comes from the enterprise having a lower cost or higher value than its competitors. There are linkages between activities within an enterprise's value chain and often the problem is one of optimising the trade-offs among these linkages. Because an enterprise's value chain is within a larger industry value system linked by value channels, it is an important focal point for identifying competitive advantage. Value chains are widely used by enterprises in many sectors of the economies of industrialised countries for identifying strategic initiatives. There are few published examples of the specific application of the concept in construction enterprises although the principles are apparent in some of the examples examined later.

Value chain analysis has much scope for application to construction enterprises. In construction, much of our concern appears to be with tactical planning. This relates mainly to the 'operations' segment of the value chain. Construction enterprises do not appear to have been as concerned with other aspects of their value providers and as such the value chain helps to focus attention on other areas that are vital to the survival of their businesses. Technology and development and human resource management are receiving some attention but procurement, sales, service and the logistics of communication with other enterprises offer more strategic scope for a multi-enterprise and project-based sector like construction.

2.2.3 Generic competitive strategies

Competitive strategies are ways in which an enterprise can analyse their strategic target either in terms of an approach which is industry-wide or aimed at a particular segment of the industry. The second approach is preferable for construction, drawing a distinction between the uniqueness of the product or service offered, in the customer's perception, and its relative cost. This two-stage analysis leads to alternative approaches that have become accepted as the generic competitive strategies of product differentiation, overall cost leadership and product focus, illustrated in Figure 2.3. There is an argument against falling between these options by being 'stuck in the middle'. Others have argued that strategies that combine these three elements can also be successful.

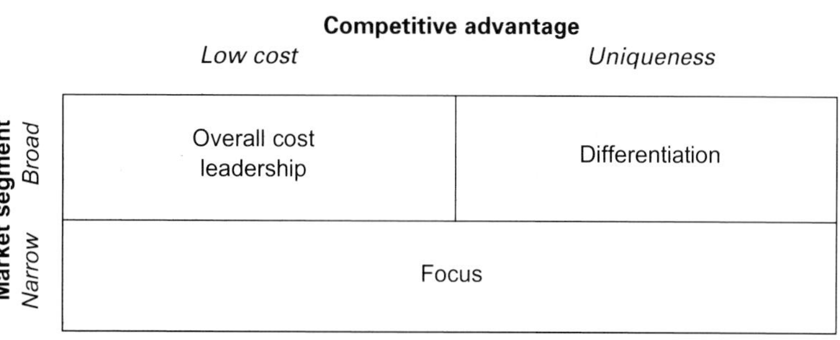

Figure 2.3 The three generic competitive strategies

There are inherent risks to enterprises following the three generic strategies and they apply in a different way to industries that are:

- fragmented;
- emerging;
- mature;
- declining; and
- global.

These approaches have been widely followed by corporate planners in many industries and in many enterprises for successful strategic management. As discussed above, despite the open bidding systems which form the basis of many construction markets, cost leadership is not the only strategic approach used for construction, especially in recent times. Concerns with quality and value for money, which, as discussed above, are on the increase, make all three relevant. There is evidence of some of these other generic strategies being followed by some construction enterprises. We will examine some examples of these later in the book.

The generic competitive strategies have clear applications for many types of construction enterprise. But you might ask what product differentiation means to a contracting company that does no design, and provides one-off projects to clients' demands. The answer must be that the enterprise should take a critical look at its strategic position in relation to markets and competitors and consider whether one or other of the generic strategies should be followed. Is it the best strategy to simply aim for cost leadership for one-off projects? Are circumstances appropriate for the company to start offering design or other services or looking for alternative relationships with clients other than one-off construction services possibly by following a partnering route supported by IT? These are the ways in which this and the other strategic management techniques can contribute, by making us change our view of construction strategy to meet the new business climate rather than dismissing new techniques because they fail to fit with traditional ways in which construction once did business.

2.2.4 The diamond model

Michael Porter has, more recently (1990), considered techniques that may be appropriate at a national and institutional level in particular. Porter's work has become widely accepted within business circles and built upon by others. It is based on a very simple view of business dynamics and competition. Porter offers an analytical framework for strategic economic management at a national level. He suggests that the economic progress of nations depends mainly on continual increases in productivity realised by groups of companies within certain industry segments that search for competitive advantage. Governments play a key role in helping these firms to upgrade, by creating a conducive environment. Professional institutions also may have potential to create a favourable environment. The features, shown in Figure 2.4 (Porter, 1990), that distinguish good environments from bad ones are:

- factor conditions – the traditional factors of production and the ways of making the best use out of them;

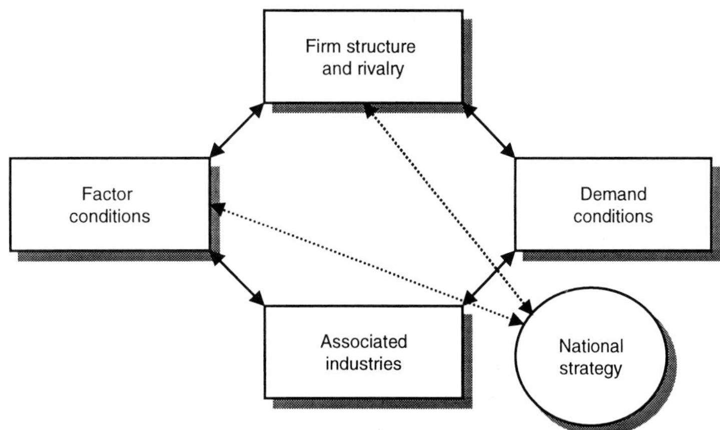

Figure 2.4 The diamond model (source: Porter, 1990)

- the strategy, structure and rivalry of domestic competition which provide companies with the experience of competition;
- demand conditions – the extent to which the nation's shopping habits challenge producers to improve quality and services;
- clusters – groupings of associated industries with common interests, where business people can encourage each other.

These four features create pressure on groups of industry segment members in some countries to invest and innovate.

Factor conditions

Taking the first of these features, factor conditions can be advantageous by virtue of having human, physical, knowledge, capital or infrastructure resources. The particular hierarchy amongst factors and the facility that exists for factor creation also may be the source of competitive advantage. Sometimes, competitive advantage can grow out of selective factor disadvantage because of the particular pressure this gives to innovation. Porter (1990) quotes the Italian steel industry and Dutch cut flowers as examples. He also shows how Korean construction contractors used their relatively low labour unit costs to exploit competitive advantage in international construction markets. Yet relative labour costs constantly change and this Korean advantage is now less pronounced.

Strategy, structure and rivalry

The nature of competition between rivals within an industry sector can vary considerably. If the competition is 'cut-throat', driving down costs and leading to poorer performance by all competitors, it is disadvantageous. Strategy, structure and rivalry can also be beneficial, by providing mutual self-support and leading to continuous improvement among its members. A good example of rivalry supporting international leadership within a sector comes from the UK auctioneers' cluster. London dominates this sector to this day because of the mutually supporting nature of inter-firm rivalry and structure.

Demand conditions

Demand conditions are the pressure that buyers bring by virtue of their tastes and expectations. These can be favourable if buyers' demands anticipate those that are later found in other countries. The size and growth patterns of demand are also important. In construction, where government is typically a major source of demand, we must take particular note of the different nature of public and private sector demand. A particularly rigorous public sector demand could give rise to a force for competitive innovation in private construction. The recent Egan Task Force report and activities of major clients in the UK Construction Round Table may emerge as a major strategic opportunity for UK construction companies, that eventually might help their strategic position in international construction markets.

Cluster

The feature of related and supporting industries is one by which an industry segment can benefit from a conducive environment. This can be through association indirectly with an advantageous industry segment that is upstream or downstream. The spin-off benefits between segments are often a major cause of superior performance.

The other component shown in Figure 2.4 is national strategy. This comes from Porter's (1990) use of government as a potential influence of the others. Our supposition here is that professional institutions and trade associations, or groups of companies, in construction can also polish the diamond in a similar way by contributing to a national strategy.

In successful industry segments within nations, all four features, which form the primary elements of the diamond shown in Figure 2.4, reinforce each other. On the other hand, in bad environments, each feature can bring about decline by hindering the efforts of individual companies, or participants in the industry segment, to upgrade. For reasons associated with these four factors, groups of companies or participants within particular industry segments in some countries have developed competitive advantages over other groups of companies from other nations where the environment created is less advantageous. Examples of this are to be found in the ten economies Porter has studied in various parts of North America, Europe and East Asia. Helping firms to increase their productivity has led to higher rates of economic growth in some countries than others.

Other industries or segments that Porter (1990) identifies as having competitive advantage due to a conducive environment include:

- Italian specialist leather footwear;
- German printing;
- Singaporean port services; and
- Korean video cassette recorders.

He then goes on to show clearly how either governments or individual participants can polish the diamond for a particular industry segment. What concerns us here are not just governments but also the professional institutions, construction enterprises, the projects on which they work and their products. These were not major players within Porter's analysis. Is there scope for them to apply this, and other strategic management techniques, in construction? What will be the effect of IT on this?

2.3 A critique of Porter's models

Many economists distinguish between 'supply side' and 'demand side' views of strategy. These terms have found popular application by macro-economists. Porter's earliest model for analysis of the potential for superior business performance in industry sectors, the value chain, concerns itself largely with production and competition with other producers. But an underlying principle of both the value chain and of the generic strategies is the need to consider both supply and demand in strategy formulation. The value chain concerns itself very much with the demand for goods and services. The generic strategies emerge from viewing a combination of source of strategic advantage (supply) and the strategic target (demand). Much of the justification of competitive advantage as a new alternative economic theory to comparative advantage (Porter, 1985) arises out of its concern with both demand and supply side issues. We would argue that both demand and supply side issues must be embraced in strategic management and that the analysis here does this.

In a book of this nature, we do not set out to provide a complete review of academic works relating to strategic management or a comprehensive analysis of the nature of competition in construction. Our objective is to outline the scope for applying some of the key strategic management tools (and not only those proposed by Porter) in construction. We use examples of successful strategies adopted by construction enterprises in different countries to highlight the merits of such systematic longer-term, corporate approaches. Many of these tools are customer-focused and service-oriented, and not wholly supply-based.

The construction industry sector has:

- a service provision nature;
- a range of enterprise types and sizes;
- a range of products offered;
- a variety of combinations of inputs as well as methods and practices; and
- a demanding customer environment, whether supply or demand led.

The industry depends on:

- the level of market development;
- enterprise size;
- service or product nature;
- the target market;
- cost range; and
- time.

Demand side issues are growing in influence and some of the alternative strategic management approaches will find increasing application alongside Porter's work. In addition, other developments in supply side thinking concerning lean production and lean thinking are being increasingly applied in construction. This new production philosophy has been observed to give order of magnitude improvements in productivity and competitiveness in the car industry. It has been advocated for wider application in other industries. If it and similar supply side developments (e.g. just-in-time and total quality management) can bring significant improvements in construction productivity, they may be seen to be of equal or greater consequence than demand side developments alone.

Indeed, we must fully appreciate the peculiar nature of construction activity. Unlike other sectors, the concepts of strategic management can be applied at five interrelated levels in construction. In this book, we critically review Porter's most recent work (1990) which is concerned with the attributes of a nation or industry segment and whether these do or can be made to promote the creation of competitive advantage by enterprises within the nation or industry segment. We argue that these concepts can be applied by construction industry institutions. We find the model useful in analysing the construction industry to highlight strategies which should be adopted by the professional institutions to support firms' and practitioners' efforts in strategic management.

The issue of the human dimension in construction management and economics may be a related, but separate, issue which reflects the richness of the concerns of our discipline. We have chosen not to focus on this here in this work on strategic management in construction although it clearly has a place. We highlight the issue of fragmentation within the construction process and project teams in later chapters of this book. In a multi-dimensional field such as construction, the issues relating to any particular matter or idea are numerous. Moreover, the absence of a recognised body of knowledge on the economics and management of construction makes it necessary for concepts to be borrowed from other fields of study. Some of these will not be universally agreed upon in their fields of origin nor would they be a complete fit for the construction industry. That should not necessarily detract from their usefulness.

2.4 Benchmarking

We cannot leave an introduction to strategic management techniques without making some mention here of benchmarking. This subject will be returned to in much greater detail in later chapters in Part B of this book. Benchmarking is objective competitive analysis. Its main purpose is to measure a company's product or service, and the business processes which support them, against:

- the company's goals and aspirations;
- its competitors; and
- best-in-class companies in other industries.

It is 'an external focus on internal activities, functions or operations in order to achieve continuous improvement' (Leibfried and McNair, 1994).

2.4.1 The process of benchmarking

Usually, benchmarking is a one-on-one activity; it is used by one company to help identify improvements in its own processes, by exchanging information with another. Normally, the activity is of mutual benefit.

You may well think that it cannot be possible to look so closely at your competitors. Yet this is precisely what is happening. It is not unknown in construction circles. In partnering arrangements, especially those where clients or customers are to share most of their business amongst a few select suppliers, the necessity of benchmarking becomes more apparent. Partnerships are implicitly about openness and communication. Furthermore, benchmarking is directed at technical or general managerial processes and so commercially sensitive information is unlikely to be exposed. Joint venture contractors are often very keen to have their own IT systems used. This exposes all their clever thinking to their partners who may also be competitors in other fields and on other projects.

Setting objectives and goals that ensure competitive excellence strikes at the heart of performance measurement. Benchmarking is the means for identifying performance levels and provides the basis for continuous improvement. For it to work successfully, benchmarking has to be:

- stakeholder driven;
- forward-looking;
- participative; and
- focused on quality.

A consequence is that it forces issues out into the open so that they can be dealt with according to the goals of the company. It identifies current practice and, through external comparison, identifies best practice and the actions which are needed to match and exceed that best practice. By bringing in the external dimension, energy can be channelled away from internal conflict towards a focus on achieving a competitive advantage.

Benchmarking encourages step change and has potential to support radical business redesign as similar processes are often conducted in completely different ways in other companies or industries.

2.4.2 Extent of application in other industry sectors

Credit for bringing benchmarking to the attention of the Western business world belongs to Xerox Corporation, through their response to the threat of Japanese competition in the late 1970s. It is widely held that, without a radical shift in thinking and behaviour, Xerox would probably not have survived, at least not in its present form. At the time, Xerox were faced with competition from Canon who were able to offer comparable copiers distributed to their markets for considerably less than it cost Xerox to just manufacture them. Like so many others, Xerox could not afford to wait a couple of decades to take the slow, but proven, path to continuous improvement trodden by the Japanese. Short-cuts that would allow them to compete and stay in business were called for. Benchmarking was the response to that need. Soon, it reduced costs by 50% and defects to levels comparable with Japanese industrial products.

Since then, benchmarking has grown to become part of the toolkit of management in companies around the world. Examples in banking, retailing, air transportation, automotive manufacture, communications and IT provide evidence that it is no isolated activity. Conditions that typically trigger the need for benchmarking include:

- quality programmes (British Airways plc);
- cost reduction/budget process (IBM);
- operations improvements (Exxon Chemical);
- new ventures (Toyota–Lexus); and
- competitive assaults/crises (Xerox).

With one exception, these are examples of reactions to events rather than a response to the need to assess the company's performance continuously. Nevertheless, they provide sufficient incentive for management to bring about a re-evaluation of business processes. The Toyota–Lexus example is noteworthy for the way in which an entirely new car was created from attention to competitors' products and proven parts, the extent of which was previously unknown in the automotive industry.

Benchmarking is not, however, something which is directed

solely at high profile or cost-significant processes; it has application almost everywhere in an organisation. Many minor business processes are subjected to study in a large number of organisations. Enough of them will make a mark on overall product or service delivery. This can mean more profit instead of unnecessary cost.

2.4.3 Necessity of intra-company and cross-industry comparisons

Benchmarking is the opportunity for companies to document and review their business processes. Internal examination may provide part of the answer, but it can never provide it all. For benchmarking to mean anything, it requires objective competitive analysis through comparison with external organisations. Comparison between companies competing in the same sector and across industries provides that objectivity. This is the rationale behind the 'external focus on internal activities'.

2.4.4 Wider benefits of benchmarking in the construction industry

For construction companies – many of which offer capacity as opposed to a product or productised service – the need for benchmarking is perhaps greater than in manufacturing. In manufacturing, goods or products can be objectively compared, one against another, along with the processes that create them. Understanding how to do things better is more readily apparent. Construction offers no such transparency and relies on procedures and practices that are often ill-defined and poorly documented. From a client and customer perspective, differentiation between construction companies is not always easy. Lowest bid price is often the only way of choosing one company over another.

Between 1992 and 1995, the price of construction work in the UK dropped by around 26%. Given that Sir Michael Latham's report, *Constructing the Team*, called for a further 30% reduction by 2000, construction companies, amongst others, would appear to have a very tough time ahead of them. Sir John Egan's Construction Task Force report (1998) has shown that the pain is continuous by calling for annual improvements of between 10 and 20% in cost, time, accidents, profitability and investment. Never was strategic management more important in UK construction.

A proportion of the saving so far can be attributed to market forces; margins have been cut all the way along the supply chain. But this cannot account for the total figure. Some structural change must have taken place. Reductions in layers of management may have been the rapid response by some companies, as a way of reducing overheads and raising competitiveness. This action is not necessarily a consequence of considered plans which strive to increase efficiency whilst lowering operating costs. Survival has played a large part, as time does not permit gradual changes to take place. Short-cuts are therefore needed and, again, this is where benchmarking has a part to play.

The marketplace for construction in the UK is likely to suffer from overcapacity for some time to come. When competition is greatest, as at this time, pressure on prices is downwards. On its own, this should be a significant reason to begin benchmarking in earnest. The traditions within the industry do, however, militate against this. Confidentiality tends to overrule most other considerations. When companies are fighting for survival, few managers are going to want to appear to be asking for their systems and processes to be opened up to competitors. Streamlined business processes would surely benefit any company - through lower overheads - and provide competitive advantage. The gains might, however, be relatively short lived. If you accept that the competition would soon catch up, then companies really do have to recognise the need for continuous improvement. This is why benchmarking is so important and the scope for its application extensive.

A further argument for construction to be considered for benchmarking is that if ever there were an industry or sector that needed objective measures (or benchmarks), this is it. The industry and its clients suffer because there are too few measures, cost being the notable exception. This absence gets in the way of improving the processes to deliver better end products. A lack of understanding of the relationship between quality and price, something that manufacturing has grasped and overcome, is all too evident. Clients and customers simply do not know what they ought to be paying for something, except as little as possible. Issues of value must come into play. Since the quality/price relationship is not properly understood, clients may be unable to judge whether or not they have achieved value for money. Benchmarking has the potential to change this position. This is why it has become a key feature of the changes called for by Egan (1998).

2.5 Other approaches to strategic management

2.5.1 Core competencies

A more recent approach suggests that a company should concentrate on progressively improving its core competencies (its key abilities and skills). NEC, the Japanese electronics group, framed a strategy in the 1970s, still being applied, which stresses the importance of honing its skills in semi-conductor making, a process which is common to its communications, computers and components businesses. This has enabled it to switch more quickly than many of its competitors into new products such as laptop computers and mobile telephones. Note that this is not the same as the approach of 'doing best what one knows best', which many construction practitioners would subscribe to. Rather, it implies identifying a particular aspect of what one does best, preferably a basic competence. It then implies striving to remain ahead of the competition in that particular area, while using the lead in that area to improve on overall operations and maintain or enhance competitive advantage by introducing relevant new goods and services. Being able to identify a core skill for an enterprise is an important part of a corporate strategy. There is scope for application of this technique in construction especially for those enterprises working in the high-technology parts of the AEC sector. The Egan report again emphasises the paucity of specialists and branding in UK construction. Stalk *et al.* (1992) develop the core competences concept further into a core capabilities strategy.

2.5.2 Strategic vision

One can go further and argue that in today's business climate there is a need for companies to have a strategic vision of the unique contribution they wish to make. The Body Shop, Sony, Benetton, Swatch, Wal-Mart and Merck are all cited as examples of companies that have successfully exploited visionary strategies. To become a visionary organisation requires a strong underlying philosophy with challenging, shorter-term goals or missions. It is difficult to demonstrate many examples of visionary construction organisations although Japanese contractors' considerations of space construction display a greater vision than most. Takenaka's plans for Sky City 1000 (a 1000-metre high building with fourteen layered plateaux of houses, shops and offices with integral monorail) are described in their annual report as one of their new horizons. Shimizu's TRY2004 pyramid city is similarly a visionary concept.

2.5.3 Keeping track of trends

The best approach to analysing a company's competitors and adopting suitable corporate plans and policies is to:

- look at a rival's existing products, however diverse, to ascertain whether there is a particular skill underlying and underpinning them;
- assess whether these skills add value to customers; and
- choose and constantly improve upon a core competence which should, preferably, be difficult for competitors to duplicate, and will thus provide a sustainable competitive advantage.

A key to this third component of the strategy is to keep track of trends and developments in the basic knowledge underlying it, for example, by visiting rivals' facilities, suppliers and university research departments. In construction, visionary publications that emerge from time to time may contribute to this.

2.6 Definitions of business process redesign

There are numerous definitions of business process redesign (BPR) which can often complicate the concept and confuse the issues. In analysis of the definitions of BPR, we will not focus on the detail of each definition, but examine the common threads running through all the definitions. This approach will help you to understand the important elements of the concept, rather than immersing you in the detail of definition. Five prominent definitions of BPR are shown below.

- Hammer and Champy (1993) – 'the fundamental rethinking and radical redesign of business processes to achieve dramatic improvements in critical, contemporary measures of performance, such as cost, quality, service and speed.'
- Davenport (1993) – 'The term process innovation encompasses the envisioning of new work strategies, the actual process design activity, and the implementation of the change in

all its complex technological, human and organisational dimensions.'

- Petrozzo and Stepper (1994) – 'Reengineering is the concurrent redesign of processes, organisations, and their supporting information systems to achieve radical improvement in time, cost, quality, and customers' regard for the company's products and services.'
- Venkatranam (1991) – '... business process redesign involving the reconfiguration of the business using IT as a central lever. Instead of treating the existing business processes as a constraint in the design of an IT infrastructure the business process itself is redesigned to maximally exploit the available IT capabilities.'
- The Boston Consulting Group (1994) – 'A senior management-led redesign to achieve breakthrough strategic impact on the business.'

2.6.1 Common element: process

Placing the process at the core of business analysis was a central feature of the quality movement developed over 40 years ago. It has been used to great effect by Japanese managers, contributing to their industrial and economic success. The development of process analysis has been significantly slower amongst Western companies, although the adoption of TQM and BPR suggests that the West is keen to centralise management philosophies around business processes. This is seen as a more effective method of combating the demands of an increasingly customer-oriented global economy.

As with the definition of BPR, the definition of a process varies amongst theorists, although the various definitions are conceptually comparable. Hammer and Champy (1993) define it as:

> **"... a structured, measured set of activities designed to produce a specified output for a particular customer or market."**

while Davenport identifies:

> **"... a collection of activities that takes one or more kinds of input and creates an output that is of value to the customer."**

The Boston Consulting Group use as a working definition:

"A sequence of activities that create value for a customer."

Processes cut across traditional functional boundaries, forcing the organisation to analyse how products or services are treated from inception to completion. Processes must have inputs that are transformed into outputs that are of value for a customer. The definition of a customer is wider than an external entity; they can also be internal within the organisation. Managers lose sight of the overall process by working in functional departments. Few companies analyse performance based on process criteria. Their business processes have evolved from *ad hoc* decisions made by functional units which are implemented to improve their own effectiveness, rather than the effectiveness of the whole process. How many construction companies are still organised with self-optimising estimating, accounting, planning, buying and contract management functional departments? BPR reverses this practice, maximising the performance of the entire organisation by focusing on the interdependencies within and across the sometimes disparate functions that comprise the organisation.

Process structure has been the subject of debate between BPR theorists. Although any company could count between 2 to 200 business processes, there will be an optimum number. Johansson *et al.* (1993) identify between five and eight core processes (value creating due to the competitive capabilities they give the company) within any industry structure, while Davenport (1993) numbers between 10 and 20. Limits are set on the number of core processes for two reasons.

- If a process is too broad, it will be impossible to fully understand its complexity and therefore difficult to substantially improve it.
- If it is too narrow, BPR will only produce minor gains to overall business performance.

Arguably the most important change any company actively engaged in BPR can make is to re-orientate organisational structure from function to process. Justification for this substantial change is that business objectives are achieved by processes, not functions. Adam Smith's concept of task decomposition broke down the process into easily manageable parts to provide a framework that maximised output. The evolution of the contemporary functional organisation can be traced from here. When these principles are applied to making pins or the Ford Model T assembly line, it is easy to understand the logic of the philosophy. But, as companies change focus from producer to customer, the new imperatives are flexibility, innovation, speed, service and quality. It would seem that business processes are the most effective way of dealing with the new competitive environment.

Process analysis in functional organisations is often characterised by a high number of 'hand-offs'. A hand-off is 'the complete relinquishing of an unfinished product by one worker (or group of workers) to another worker (or group)'.

A hand-off usually means that time is lost and accountability shifted. This lengthens cycle times and ultimately causes inefficient utilisation of resources. While a number of hand-offs are essential (especially in a construction environment), all hand-offs should be examined with a view towards elimination.

> "Frequently the remnants of functional processes invented in the 1950s, hand-offs can be the most insidious problems in a process."

The IBM Credit case study in Chapter 4 highlights the potential of the removal of hand-offs to facilitate process effectiveness.

As more people work in less permanently structured organisations, in the aftermath of the information revolution, the need for whole processes to be owned becomes greater as participants in making finished products may never meet each other. They will more typically work in different companies and continents.

2.6.2 Common element: redesign

The approach of process redesign has been subject to some debate. Hammer and Champy (1993) label the philosophy necessary for successful BPR as discontinuous thinking, defining it as:

> "...identifying and abandoning the outdated rules and fundamental assumptions that underlie current business operations."

Discontinuous thinking implies a clean slate or fresh approach to process redesign to harness current knowledge and process improvements to key performance measures that are radical.

In researching over 100 companies Hall *et al.* (1993) introduced two key concepts essential to facilitate successful BPR. These are 'breadth' and 'depth'.

- **Breadth:** identifies activities that directly create value within a process. A narrow breadth would focus upon a sub-process or even a single task, while a wide breadth involves the redesign of broad processes that are critical to the business. The research found that companies who ambitiously apply BPR to core processes were high achievers. A broad agenda also examines interdependent activities, hence reducing hand-offs and communication flows.
- **Depth:** defines the depth that the BPR initiative has to penetrate into the organisation to effect lasting change. They identify six depth levers that all need to be fundamentally changed to ensure escape from the *status quo* of traditional organisation practices. These are:
  - roles and responsibilities;
  - measurements and incentives;
  - organisational structure;
  - information technology;
  - shared values; and
  - skills.

Some observers have questioned the need for radical change in the approach to redesign. Due to the high risk of BPR failure, it should only be introduced if traditional incremental improvement measures have failed to provide the company with the performance improvements required to remain competitive.

2.6.3 Common element: information technology

The impact of IT on business has not yet yielded the productivity increases its pundits promised – the adoption of process methodologies aims to address fundamental flaws with the application of IT. The main objectives for information systems have been automation and process rationalisation which have only produced minor incremental productivity improvements. BPR protagonists see IT-induced process redesign as its saviour. The belief is that by changing the way we work and taking advantage of IT's potential to deliver competitive advantage (Porter and Millar, 1985) we will see business fortunes increase dramatically rather than incrementally.

IT has not been incorporated into many of the BPR definitions, but it is central to all of the methodologies. The effective use of IT as an essential corporate infrastructure undoubtedly has enormous potential to change the way we work. IT-leveraged BPR is where processes are redesigned that utilise the available capabilities of IT, instead of designing processes and using IT to speed them up.

Although the importance of IT to BPR is undoubted, the evolution of our understanding of IT in a process environment is at an early stage. Our articulation of the interconnection between IT and business strategy is also at the initial stages, even though we are clearly observing the emergence of fundamental linkages between strategic management and IT.

2.6.4 Common element: strategic management

We need to set strategic management in the context of BPR. Strategic management only fits into one of the definitions of BPR, but is implicitly recognised in all the methodologies. To be successful, BPR must be strategically led. The design of new business processes is defined by the strategic plan. Without the guidance of strategic planning, BPR will not have the focus to be successful and it is likely that different processes will have different aims and objectives and be incongruent.

Although it is expected that BPR is strategically led, BPR and strategic planning have a recursive relationship. BPR offers new capabilities and skills that are exploitable at the strategic level. The strategic planning process should therefore account for increased organisational capabilities, not planning within the confines of the pre-BPR organisation. These issues are developed later in the book.

BPR includes the wider concepts of 'business network redesign' (BNR) and 'business scope redefinition' (BSR). BNR calls for BPR principles to be applied to supplier and customer partnerships with the aim of adding value to mutually beneficial relationships. BSR addresses the enlargement of the scope of the business and the shift of business scope, in the ability of companies to enter new markets. The example of Otis Elevators in Section 4.3.3 highlights this. Future market leaders will create competitive advantage through the creation and exploitation of emerging markets.

2.7 Summary

This chapter has laid further foundations for much of the remainder of the book. We have seen that there are a wide variety of viewpoints and perceptions about strategic management but that there are some commonly recurring themes. These include the need to take a long-term view, to be customer-focused and to search for competitive advantage. A number of needs have been identified, for example, the need to look towards diversification and new ways of doing business and the need to put the business in control of its destiny. The techniques and models of Michael Porter, whilst being subject to some criticism, undoubtedly have wide application and will be referred to in later chapters. We examined the technique of benchmarking and its possible applications in construction, together with a variety of other alternative approaches to strategic management. Finally, we saw how business process redesign could be used as a strategic management tool by exploiting the real capability of IT to improve business performance.

3 Strategic management in construction

Martin Betts and George Ofori

"Within industry, our need to innovate and look fundamentally to the future, has never been greater."
– Chief executive, major UK consulting group

The techniques of strategic planning and management apply in a unique way in the construction sector. The sector operates at a number of different levels. The way strategic management works at these different levels and the constraints and influencing factors at each level are complex.

Strategic management is applied in construction at present but not in as extensive or formal a way as in other industries. There is much benefit to be gained from more extensive strategic management in construction companies, as some specific examples can demonstrate.

- Strategic management in the construction sector applies at the level of the national economy, the professional institution, the company, the project and major buildings or products.
- Strategies of construction companies embrace integrated approaches, differentiation, diversification and core competencies.
- Current approaches to strategic management in construction are rather ill-applied and can be improved.
- Some construction companies have obtained substantial competitive advantages from following improved strategic management processes.

"The construction industry has been working on inappropriate, and thus inefficient, systems for years. Most managers no longer see any point in investing in information management systems."
– Corporate purchasing manager, French contractor

3.1 Introduction

We have now defined the principles of strategic management and process analysis and shown why they are important. This chapter takes us further by exploring how these techniques can be applied in construction. We will see that the principles can be applied at a number of different operating levels. Examples of the way IT is used in construction at different levels will follow in later chapters. At this stage, it is important for the reader to appreciate the scope for the techniques within our sector.

Given the level of risks involved in construction activities and the turbulent environment in which they operate, you might think construction firms would have been undertaking strategic management for a long time, simply to survive. Within large UK firms, few seem to be explicitly aware of strategic management concepts, and many firms have no written strategy. However, when asked about strategy, senior managers are able to talk enthusiastically and cogently about it. Owing to the uniqueness of projects, management of a construction firm is essentially:

> "management of change, coping with changes in the environment and making adjustments to strategy, diversifying as necessary, modifying ... operations, altering methods of employment of manpower, updating the approach to managers, changing organisational structure and making constant adjustments to financial and pricing policies."

Construction enterprises have adopted appropriate organisation structures, practices and strategies to suit the nature of their tasks and their peculiar (traditionally turbulent) operating environment. The industry's approaches and practices have useful lessons for other sectors of the economy facing uncertain environments.

On the other hand, the inability of construction enterprises to develop appropriate responses to such issues as:

- the changes in levels of activity;
- fluctuations in demands for personnel; and
- the relatively low productivity associated with construction work,

indicates that the industry lags behind others in systematic strategic management, in seeking niches, and in identifying and improving upon core skills.

As we have seen in Chapter 2, a business process perspective of construction is also less often adopted than in other industries. Construction is typically seen as a project-based, problem-solving activity in which short-termism dominates. Developing a process view of business is relatively new to construction but recent indications show that the larger companies are undertaking process mapping and improvement exercises.

3.2 The rationale for a five-level framework

Strategic management and business process analysis are clearly different issues when applied to construction than they are in other industries. The complexity of construction is considerable due to its structure, its historical evolution and the nature of its process and product. It is therefore necessary to consider several levels at which strategic management and process analysis and positioning may be applied, as shown in Table 3.1.

Table 3.1 Five-level framework for strategic management in construction

Levels of application		Responsibility for implementation
Level 1	National construction industry	Public-sector agencies
Level 2	Professional institution	Professional bodies and trade associations
Level 3	Construction enterprise	Enterprise
Level 4	Construction project	Client and project team
Level 5	Construction product	Client and project team

Nations, and their construction industries, need to adopt strategies through their public agencies for a number of reasons, including:

- the globalisation of construction;
- the pressure on resources on one hand as against rising expectations of quality on the other; and
- the need to adopt a long-term perspective to appropriately position the economy.

Responding to the Latham review of the UK construction industry

(1994) and the more recent Egan report (1998), the idea of a national construction strategy is being considered in the UK at present by groups such as the Construction Industry Board (CIB) and the Construction Research and Innovation Strategy Panel (CRISP). A process basis underlies many of these initiatives. At the very least, a national-level strategy would provide the framework, or set an example, for corporate initiatives. Such a strategy would also ensure that measures which support the activities of the enterprises are planned for and implemented.

Many construction industries are influenced by their historical development and the professional institutions that have resulted from it. Although major deregulation, as faced by other sectors, appears inevitable, there is still scope for strategic management and process analysis by the existing professional institutions and trade associations. This is because of the power and influence the professional institutions retain in some parts of the world, the expertise they can muster and the scale and scope economies which can be derived from their strategic initiatives. The professional institutions have much scope to analyse their members' processes and to influence the way they are conducted through their education and membership roles.

The most significant (and common) examples of strategic management and process analysis in construction occur at the corporate or construction enterprise level. Indeed, most of the strategic management concepts and techniques which are presented in this book are meant to be applied by business organisations. Much of the new thinking in business process analysis also most obviously addresses the corporate or enterprise level.

In this book we will mainly consider corporate applications but it is important for us to consider the peculiar nature of the construction process and product. In contrast to business and manufacturing industries, the fundamental operating level of the construction industry is the project. It is reasonable to assume that the project is a level at which business strategy can be applied. Many major innovations typically occur first within major projects. The large Japanese firms, in particular, have been successful in applying business strategies in relation to projects based on the concept of the integrated architecture, engineering and construction (AEC) enterprise. In other parts of the world, notably the UK, this is an area where innovative clients such as BAA, Tesco, McDonald's, Marks & Spencer and Sainsbury have sought strategic advantage.

Having outlined five levels by which to examine opportunities for strategic management and business process analysis in construction, we can now consider the factors giving impetus to the need for strategic management and process analysis at each level, and some examples from around the world are described in overview.

3.3 The industry level

One of the earliest examples of a national construction strategy can be found with the Singapore Construction Industry Development Board which formulated a strategic plan for the construction industry of Singapore to develop it into:

> "...one which is not only capable of achieving timely delivery of its final product, but is able to do it at increasingly higher levels of attainment. It must be an industry which affords the man in the street more choice to customise his constructed habitat. Equally important, it must be one which can seize and exploit business opportunities on a global basis."

The plan had the following basic components.

- Cost competitiveness – a study compared the Singapore construction industry with those of other countries, resulting in schemes to improve construction management, promotion of R&D, further use of modular co-ordination and prefabrication, and application of information technology.
- Quality – introduction of quality management systems and structured quality assessment, combined with measures for improvement of management and operative skills.
- Export performance – encouragement of strategic alliances and consortia among firms for export, and provision of export-market intelligence, technical advice and incentives.
- Strategic sectors – development of capability in refurbishment, intelligent buildings, high-technology production facilities and advanced engineering works.

More recently, the Singapore Ministry of Trade and Industry sponsored a major long-term economic planning exercise for all sectors. Their *Vision 2030* foresees Singapore per capita standards of living rising to those of the US under a high growth scenario. The Strategic Economic Plan (SEP) also envisages initiatives for:

- national teamwork;
- international orientation;
- innovation;
- industry clusters;
- economic redevelopment of low productivity sectors;
- international competitiveness; and
- reducing vulnerability.

Specific ministries were allocated tasks within these initiatives. The construction sector features in the SEP with a vision having been set of enhancing Singapore's competitiveness in construction services and establishing itself as a centre for services for the Asia Pacific region. Within the plan, the construction cluster was allocated a number of goals that are needed to enable the SEP objectives to be met.

The work by the Singapore government identifies clusters where a good combination is found of parts of Porter's 'diamond' strategic management technique which we examined in Chapter 2. The construction cluster is seen to have the key industry drivers of:

- demand for commercial, industrial, public housing and infrastructure construction; and
- growth in both South East Asian and developing countries.

The core activities of the cluster, or those where the diamond model shows a competitive group of companies, are seen as being those of private sector developers and contractors, with their export of construction services being of key importance. Architectural, engineering and specialist design and consultancy services are seen as secondary activities within the cluster. Arising from analysis of the cluster, the four key capabilities which the Singapore construction industry now plans to develop are in relation to: skilled manpower, building technology, the scope of professional and construction services, and a good track record.

Using the example of Singapore, we can see that a number of the concepts and techniques of strategic management and process analysis are appropriate to the level of the national construction industry. At a more general level, Singapore had been graded the most competitive national economy by the World Economic Competitiveness Report for three consecutive years between 1996 and 1998. With regard to other countries, the Japanese Ministry of Construction has been active in strategic management at an industry level and the UK government is active in attempting to strategically manage UK construction through the CRISP initiative and the CIB. Within New South Wales (NSW) in Australia, the Public Works and Services Dpeartment, which spends A$6 billion annually on construction projects, has devised a public construction strategy that it has called Construct NSW.

3.4 The professional level

To establish whether the professional institutions and trade associations can use, and are using, these new concepts and techniques, we need to look at the current nature of strategic management and process analysis in construction at the institutional level. We will examine the general strategies of professional institutions and trade associations, and the context in which they operate.

3.4.1 Influencing factors

The need for strategic management and process analysis at the professional institutional level is almost peculiar to construction. It is in construction that the people and organisations with different specialisations have recently been in almost constant competition to extend their respective share of total activity. Whereas each profession generally recognises the technical competence of others, the competition is for the management/coordinating role on the project. As the manager/coordinator is the leader of the project team, the stakes are high. Indeed, since various countries organise their construction industries differently, sometimes without involving professionals who are traditionally important in other countries, the competition is, on occasion, based on technical competence, or claims to it. The relevance of the skills of a particular profession is sometimes called into question by members of others. The quantity surveying profession in Tanzania faced such a fate in the early 1980s. Arguing that the quantity surveyor made no contribution to the project, the other construction professionals launched a concerted effort to demand the banning of the quantity surveying profession (which is accorded official recognition and protection by requiring persons to be duly registered before practising). They were unsuccessful in their attempt but this example demonstrates the level of competition amongst professionals and professions.

With the introduction of new systems for procuring construction projects, the jockeying for position amongst the professions to highlight their suitability for the leadership role has intensified. Another factor giving impetus to this competition results from changes in the statutory protection offered in some countries to certain professions in all, or certain aspects of, their work. It is also unlikely that professions which do not already enjoy such protection, and have been calling for it, will be accorded such a favour in the future. These are aspects of the five-forces model affecting the construction professions and the strategy, structure and rivalry aspect of the diamond model, which give particular potential for competitive advantage from timely strategic positioning.

The strategies that UK professionals followed in the past were re-oriented to face dynamic changes within Europe. This was because of the different range of construction professions in member countries, the different roles played by professionals on projects in each country, and the varieties in the relationships among them. All EU countries have architects and engineers although their role varies. Only the UK and Ireland have professional surveyors and builders. These differences are determined by historical reasons but often, factor and demand conditions lead to them remaining in place. Both would have to be changed to create the new environment for professional demarcations. Many changes within the EU market for construction influence these factor and demand conditions. The professional institutions may be able to directly influence them further. The mutual recognition and harmonisation of qualifications will begin to affect factor conditions. The greater dispersion of other economic activities to a European basis will influence demand conditions. As we saw in Chapter 2, Porter argues that it is during times when the four attributes of the diamond model are changing that organisations have the greatest scope to reposition themselves in a way to gain competitive advantage. This is still the case for all European construction professions, with both opportunities and threats present. Professional institutions influencing the business processes conducted by their members is a major mechanism for doing this.

3.4.2 Responses

In several countries, professional bodies and trade associations have endeavoured to equip their members with the necessary skills and facilities to enable them to face the challenging future with confidence. Skills are developed through accreditation of courses, continuing professional development and dissemination of relevant state-of-the-art information. Facilities are provided through central data services, computerised libraries and the like. The intention is to help members enhance their competitiveness to reinforce the importance of their profession. Institutions also run publicity programmes to raise public awareness (or to counter adverse publicity from other quarters). In the UK, there was a prolonged (and acrimonious) debate in the 1970s and 1980s about the merits of some new forms of project procurement, and the professional best suited to play the leading or coordinating role on projects. This was followed by a response from the British Property Federation (BPF), a large client association posing threats to many professions. All were quick to react. This typified the previous (negative) approach when the institutions tended to delineate their areas of competence and guard them jealously. They resisted new thinking for several years until it was clear that such opposition was futile.

A more forward-looking approach is now evident in some quarters, while the claims to a leadership role remain in others. Introspective studies identifying new roles for particular professions have become common. The stimuli for these have come from:

- perceived threats to the profession from outside;
- new, beneficial knowledge that should be applied – such as IT;
- initiatives by client groups dissatisfied with the services they are getting – such as the BPF and continued with the formation of the Client Liaison Group, the Construction Clients Forum and the Construction Round Table, and the major thrust towards clients within Latham and Egan;
- issues of wide public concern – for example, the environment and the drive towards sustainable construction;
- concerns expressed by users about the quality of buildings and works – such as the issue of 'healthy buildings';
- geopolitical and economic events; and
- great new opportunities – such as the colonisation of space.

Loose, strategic alliances among members of associations have, in some cases, been institutionalised. These include 'arranged' tendering systems and guarantees of members' work overseas.

The response to the increasing concern about conserving resources and protecting the environment may also be cited. The pressure has come from widening statutory control, effective action by pressure groups, greater interest by clients and users, and mounting awareness on the part of the general public. In the UK, many of the professional bodies have prepared 'green papers'. However, there is scope for further action for the benefit of the individual professions as well as for the common good.

3.4.3 Strategic management at the institutional level

We can clearly see that strategic management and process analysis are appropriate at the institutional level in construction. Given that responses to threats constitute a prime motivator for strategic management and process analysis, the application of the techniques at this level is likely to continue to be important in response to:

- the pressures of deregulation;
- insistence by clients on better quality of services;
- competition from other professions; and
- inevitable threats from outside the industry.

On the basis of the review and discussion so far in this chapter, at the institutional level of the construction sector now, the use of the new strategic management and process analysis techniques and the scope for applying them further can be summarised as follows. We have seen examples of the five-forces model and of core competencies being used. The new approaches developed by Michael Porter (1990) for exploring the competitive advantage of a nation's industry segments seem to apply to professional institutions to a large extent. The principles of business process analysis and of re-engineering in particular appear less relevant to professional institutions, although in a process environment the roles and responsibilities of professionals will undergo some considerable change. The move towards the multi-disciplinary practice, and large construction consultancies serving whole project life cycles, is an indication that professionals will offer multi-disciplinary services, and it is clear from the BPR paradigm that such businesses will be increasingly engaged in partnering arrangements that would hopefully secure long-term relationships.

What we now need is a means of finding ways to extend our use of the new strategic management and process analysis concepts at the professional institution level. The job of the professional institutions could be to offer the expertise and knowledge to equip professionals with the skills to work under partnering arrangements in a process environment. They also have a key role in accepting and facilitating these structural changes that increasingly appear inevitable.

There is often a wide gap between strategic management and process analysis at the corporate level and that undertaken at the level of the professional body or trade association - what an institution perceives and offers as strategic directions for its members and what the members can apply.

For example, despite the attempts by the Royal Institution of Chartered Surveyors (RICS) to highlight new opportunities and portray a modern image of the profession, many of its members still adopt a conservative approach, and retain an old perception of themselves. Traditionally, the quantity surveying profession was identified with the construction process but now is more closely associated with the construction product: its development, its procurement, and its maintenance and management. Whereas the technologies and techniques it used, and the relative importance of the profession, had changed (for example, quantity surveyors were providing management-oriented services concerning value and procurement), the self-image of the profession was outdated. Many quantity surveyors considered themselves as:

> "... a rather technically based profession providing a ... backup, back-room service supporting the activities of the far more important skills applied within the design and management of construction" (McDonagh, 1991)

This message of a conservative profession that needs to be proactive is reinforced by the recent RICS QS Think Tank Report *The Challenge of Change* (1998). The institutions need to detect and identify the threats and opportunities in advance. They can then develop appropriate strategies incorporating measures to be implemented at the level of the institute, while offering guidelines for the firms. Each firm of professionals would then have to adapt the broad guidelines to carve niches for itself, and benefit from the macro-level information services, hardware and training programmes offered by the institutions.

3.5 The enterprise level

Having seen examples of the way strategic management and process analysis can apply at the national and institutional levels in construction, we can now contrast this with the picture at the enterprise level.

3.5.1 Definition of the construction enterprise

The term 'construction enterprise' is being used in this book deliberately. It refers to any business entity involved in any aspect of the construction process. Thus, it encompasses much more than a 'contractor' or 'building company'. The review that follows and the techniques outlined are relevant to many types of business organisation in the construction sector including general contracting firms, specialist contractors, architectural or engineering design partnerships, cost consultancy practices and development companies. It is also likely that the trends outlined above, and the application of the techniques and principles described in this book, may lead to fundamental changes in industry organisation and in the types of enterprise to be found in the construction sector. Thus, it is more appropriate to consider all enterprises within the architecture, engineering, construction (AEC) sector.

3.5.2 Relevance to all enterprises

It is sometimes suggested that concepts of strategic management and process analysis such as those described in this book are relevant only to the fields of manufacturing and services where the rate of change of technology is particularly high, and where competitive advantage can be obtained from seeking to be nimble and finding ways of assessing and out-manoeuvring competitors. Others might feel that, even if strategic management and process analysis were relevant to construction, only the larger firms should pursue it, as many of them are already doing. However, although most of the examples in this book are of strategic management and process analysis by large construction enterprises, the level of competition in construction (especially in the medium and small project categories) is no less intense than that in any sector of the economy.

Judging by the high attrition rate and the traditionally low profit margins, there is little doubt that the industry's operating environment is even more fluid than that of other sectors. In the industrialised countries, even the smallest construction firm faces potential challenges from trends such as the increasing importance of do-it-yourself among owner-occupiers of houses. General small contractors might also be under threat from the spread of labour-only subcontracting. Finally, as trade barriers come down, construction enterprises in each country will face real competition from firms from other countries, even for small construction projects, if only in the districts close to the national borders.

In the developing countries, the large construction firms will need to undertake strategic management and process analysis if they are to survive the expected onslaught of foreign construction enterprises following the adoption of free-market economic policies by most governments. Small companies will also require longer-term perspectives if they are to survive:

- downward 'plundering' by larger firms;
- continual change in public sector development budgets;
- rising client aspirations; and
- changing industry practices.

We will see that the new strategic management and process analysis techniques have relevance to construction enterprises. It is now important for us to see where these new techniques stand in relation to:

- previous considerations of strategic management and process analysis in construction; and
- current corporate management practices in construction.

3.5.3 Current corporate management in construction

Before considering ways in which the new strategic management and process analysis techniques can be applied, we should take note of the present level and nature of corporate planning in construction. The examples quoted here are taken from previous publications but later in this book we include industrial case studies from the UK and other international construction enterprises. Some of these examples of strategic decision-making were observable even within the 'traditional' structure of the industry. Through deliberate policy decisions, some contracting firms (although 'conventional' in all other respects) have been able to sufficiently differentiate their services to make them 'near-automatic' candidates for the shortlist of many a selective-tender project. In many cases, organisations have diversified themselves into integrated AEC firms and others have ventured outside construction. 'Business development' has become an important part of the operations of the construction enterprise. Competition on the basis of far more than price is becoming increasingly important. Redesigned business processes are a key mechanism for achieving this.

3.5.4 Integrated approaches

Arguably, some of the best illustrations of strategic management and process analysis in construction come from Japan, despite their recent difficulties with national and regional recession. The large Japanese contractors have successfully out-thought construction firms in many markets in various parts of the world because of the attention they give to business strategy. Japanese construction firms take a high degree of interest in long-term R&D. The possession of advanced and special technologies has been an important tool for competition for these firms both at home and abroad.

The business strategies of Shimizu and other leading Japanese contractors suggest that the current competitive forces at play in construction include:

- the level of domestic and international competition; and
- the threat of new entrants to the industry through diversification of non-construction companies.

Three approaches to strategy formulation for contractors are apparent:

- product diversification;
- business diversification; and
- market segmentation.

We can differentiate between strategies of diversification and generalisation and between offensive and defensive strategies. There are six strategies that Japanese enterprises have followed to enable them to be strategically placed for the changes in the global construction industry. These include: a trans-national approach; new business development; application of the concept of an integrated engineering constructor; adoption of the idea of contractors exploiting opportunities for total project development; technology development; and exploiting financial strategies.

A Chartered Institute of Building (CIOB) mission to Japan found that Japanese R&D is enshrined within the principle of *kaizen*, or continuous improvement. Rather than accepting bad practices, the influence of this dominant culture plays a fundamental role in the ability of Japanese contractors to continually improve and adjust to market conditions. While this approach is less dramatic in its scope than BPR, it has substantially helped Japanese contractors to maintain a competitive advantage. This is despite the origins of continuous improvement lying with an American. A culture of continuous improvement should be aligned with BPR to maintain a flow of fresh thought and innovation.

3.5.5 Product differentiation

There are four means of product differentiation in construction:

(1) offering a range of project management methods;
(2) extending from construction into design;
(3) extending into financial packaging; and
(4) extending forward in the value chain into commissioning and facilities management.

Most UK contractors until now have adopted the first two approaches, many seek to increase their involvement in financial packages particularly through PFI schemes, and some have been prepared to undertake the fourth and are acquiring a reputation for doing so. The operations of the Japanese contractors show that firms have great scope to deepen their product differentiation under each of these categories, and that the viable options may be limited only by lack of imagination. The ability to pre-sell part of a major commercial development to an anchor tenant was a key factor which clinched the tender for a Japanese contractor in South East Asia. Offering joint-venture deals (with the client) for large private-sector projects has also been a winning formula recently.

Like the other South Korean families of companies, known as *chaebols*, Hyundai has a significant involvement in construction, which accounts for about 25 per cent of its sales and around 30 per cent of profits in a typical year. It was the world's seventh largest international contractor in 1985, undertaking almost US$2 billion worth of foreign contracts in that year, which represented almost 80% of its total construction work. The firm maintained its share of global construction work despite a sharp decline in overall volumes during the 1980s by adopting a group strategy of high investment in R&D, and the strategic targeting of the 'sophisticated upmarket'.

The giant Bechtel group of the USA was the third largest international contractor in 1985 with US$3.6 billion worth of contracts representing 50% of its total volume of work (ENR, 1986). Its response to a world-wide decline in mega-projects (especially power-related projects) in the mid-1980s was to re-organise its business to emphasise advanced technology and R&D. At the same time, it scaled back the number of divisions engaged in power generation and oil industry projects. These two examples demonstrate a clear goal of product differentiation within a tightening global market and in response to identified planning issues of a domestic and international nature.

3.5.6 Diversification

Hillebrandt and Cannon (1990) provided one of the most extensive descriptions of diversification in their study of 20 UK construction firms. The contractors see diversification as a major contributor to corporate growth that was evident in all enterprises studied. The work also established that increased operational and financial efficiency and greater security were prime motivators. UK firms had moved into development, consultancy, plant hire, mechanical and electrical engineering, production or sale of building materials and components, and such unrelated areas as health care, printing and waste disposal.

Diversification had been the traditional route for growth adopted by European construction companies. Naturally, diversification has taken many forms, and has been carried to different extents in various firms. Some construction companies have integrated vertically, both upwards and downwards. Tarmac in the UK concentrated on the development and production of road construction materials and other British firms like Laing and Wimpey went into property development. Alfred McAlpine withdrew from competitive tendering for building projects in the mid-1990s and focused on special projects. Fosca and Construcciones y Contratas of Spain entered into waste management (ranging from street cleaning, through rubbish collection, to water purification). Buoygues of France diversified both within construction, with, among other things, the formation of a joint venture with IBM to develop intelligent buildings, and outside construction into television, films and various forms of manufacturing and communications in a cellular phone business.

3.5.7 Basic core competencies

Another example may be cited of a strategic approach to business by a construction firm, this time from a small company. Esco Scientific of Singapore defined its core competence as cutting-edge computer application to the design of mechanical and electrical engineering services and is one of the few construction companies in South East Asia (of any size) with a significant R&D budget. Its strategy is to continue to improve upon this basic competence, maintaining a gap between it and the competition and actively and continuously identifying niches to exploit. It has established itself in energy auditing and management, together with the design and construction of services in advanced-technology production facilities such as clean rooms. Despite its small size, this strategy has enabled it to win contracts in the highly competitive US market. The concept of basic core competencies and core capabilities has particular scope for construction enterprises other than contractors, such as consulting organisations and specialist firms.

A further example of a core competence is the use of modelling expertise by Dillingham Construction in California, which features in a case study in Chapter 5. Dillingham have made strategic use of modelling in a niche application to a new market being created by local environmental and regulatory changes.

The general experience in construction is of strategic management and process analysis being a low-profile activity that faces significant restrictions at many levels. However, it is becoming clear that the forces for strategic change in other economic sectors are equally applicable to construction. It is clear that before the changes in business climate outlined above, there was already a significant gap between the extent of such planning in construction and other sectors. This has now widened.

The techniques of strategic management and process analysis are not mutually exclusive options which should be tried one at a time. Moreover, their purpose is to enhance the firm's adaptability to change rather than offer immunity from the implications of change. For the construction enterprise, formulating a strategy should not be approached from the same viewpoint as a lottery. Nor is it a matter of coming up with something new every now and then. The important issue is to develop a long-term view based on the company's core competencies, which should be continuously improved over time to offer the necessary competitive advantages and seek new niches to exploit.

Strategic management and process analysis are relevant to all construction enterprises, large or small, regardless of the aspect of construction they are involved in. It should be recognised as an important aspect of the firm's overall activity which requires as much attention as its routine operations. It should be approached in a structured and systematic manner. The increasing body of work, not only on the techniques of strategic management and process analysis, but also relevant corporate approaches to their direct application, should be examined continually by construction enterprises to identify potentially beneficial elements.

The responsibility for strategic management and process analysis should be specifically assigned to a person or group within the organisation. To this end, the variability of enterprise size must be recognised. For large international constructors, there is a clear and growing need for strategic management and process analysis teams upon whom the long-term survival of the enterprise may depend. Many larger enterprises already have such teams. Where they do not yet exist, the task of strategic management may be assigned to the 'business development' division of a company, firmly supported by senior management. It should be staffed by multi-disciplinary members, incorporating construction professionals as well as specialists in corporate strategic management and process analysis. It must be placed and perceived as central to current and future company plans rather than as a peripheral service department. The department should apply strategic management and process analysis concepts in detail to the specifics of their current position and the resources of the company, giving due regard to corporate objectives. Many innovative strategic initiatives will emerge in the construction industries of many countries in the immediate future if such departments become widespread. The consequences of this will be restructuring, the creation of new businesses and the emergence of new products and services.

There are many small and medium-sized construction enterprises for whom strategic management and process analysis departments may not be sustainable. This does not mean that strategic management and process analysis is irrelevant to such firms. Articulating their vision for the company and formulating a set of objectives and a plan for achieving them would be a useful exercise for the proprietors of such firms. For owners of firms in this group who are unable to do so by themselves, consultants may be engaged to offer such services, as is shown in the detailed case study that follows. Small firms in developing countries would benefit from central assistance (by a contractor-development or management-development agency, or a committee set up or consultants engaged by the contractors' association) to formulate corporate strategies using the emerging techniques.

It is important to remember that enterprises outside the traditional construction industry (such as materials manufacturers) may consider construction activity within their strategic sights as they seek to integrate vertically, diversify their activities, or generally find new business. Moreover, as deregulation strips the construction professions of the protection given by legislation in some countries, the trend of firms and individuals from outside the industry (such as accountants and management consultants) offering construction-related services will be accelerated. The need for construction enterprises of all types to view strategic management and process analysis as a vital activity cannot be overemphasised. The strategic management and process analysis initiatives of construction enterprises can be stimulated by and benefit from the efforts of public agencies, professional bodies and trade associations.

Case study: SAVANT

This first case study in this chapter is a very detailed description of one particular enterprise. We have seen that a shift towards strategic management and process analysis is by no means fully demonstrated in the construction sector, although there is evidence of these techniques being applied by some of the larger enterprises. Yet, the principal organisational unit of the construction sector in most countries is the small to medium sized enterprise (SME). It is at this level where the greatest strategic scope exists, by virtue of the number of organisations. In the UK and Australia, there are more than 100 000 construction firms with an average number of employees of less than five.

It is this structure which is the industry's greatest strategic flexibility and its greatest weakness, fragmentation. It is also at this level where you would expect strategic management and process analysis applications to be most different from those found in other sectors. The enterprise level, and SMEs in particular, should therefore be able to demonstrate the unique features of the construction sector and its influence on strategic management and process analysis applications.

One problem is a lack of detailed evidence of how strategic management and process analysis can be applied at the SME level of the construction sector. We can operationalise this problem by posing the following question:

"Are there opportunities for strategic management for a sustainable competitive advantage (SCA) at the SME level and if so what characterises the way this occurs in the construction sector?"

It is this question that we attempt to answer here by means of a detailed case study. We can refine the question by expressing it in two ways. First, is there scope for SCA by SMEs in construction? Second, do the full range of strategic management techniques apply at this level?

The case study

This case study, of strategic management in Australia, illustrates many of the principles already described. We can use it to examine the way that strategic management is occurring in construction. We will also evaluate the extent of success and benefits of such innovation.

The case relates to a small project management consultancy, operating in Adelaide, Australia. It is an SME that thinks carefully about strategic issues and future consequences in what is a notoriously short-sighted industry, so the firm is certainly exceptional.

At the time (1992) that contact was made with the enterprise, it was embarking on a strategic management initiative with a leading international management consulting organisation. The case study is therefore an incidental observation of strategic management activities.

History of SAVANT

The name of the company is SAVANT. This word is defined by the *Shorter Oxford English Dictionary* as 'a man of learning or science' and more particularly 'a person professionally engaged in learned or scientific research'. This conveys much of the image that the enterprise wishes to present, and its unusual nature in an otherwise practical and experience based industry. The enterprise was formed in 1986 and was therefore still relatively young. The original business aim was to provide professional, practical project management and consulting services for project initiators in government, institutional and commercial sectors. It is privately owned by its founding principals and professionally managed by a board of directors with wide experience in capital works, business and administration. The enterprise retains its base in Adelaide but is involved in key projects around Australia. These include significant developments in Melbourne and an expanding operation in Perth. At the time of the case study in 1992, SAVANT were finalising a business plan to open a second office in Perth. This was to be through a strategic alliance with a Western Australian independent project manager.

The current range of activities offered includes:

- management consulting:
  - advice on organisational structures and project reporting systems;
  - market surveys;
  - feasibility studies;
  - site identification;
  - financial sensitivity analyses;
- project management consulting:
  - client briefing;
  - consultant selection;
  - project information systems;
  - programming;
  - financial control;
  - quality management;
  - value management;
  - life cycle costing;
  - commissioning;
- construction consulting:
  - resource studies;
  - value engineering;
  - specification reviews;
  - technology reviews;
- commercial services:
  - claims control and advice;
  - contract documentation;
  - acting as expert witnesses; and
- IT based services.

The dynamic contexts in which the company saw its activities were:

- an approaching property glut in Australia and some other parts of the Asia Pacific region;
- increasing attention being given to asset management by asset owners whose portfolios were declining and fragmenting;
- quality and value management becoming more important to clients; and
- communications within the region overcoming many of the historical problems faced by and between Australia and South East Asia.

The company foresaw a growing role for a total systematic approach to total project management and expected arrangements like build, operate and transfer (BOT) to become more common. Aware of this dynamic context and through experiencing a decline in workload for its current activities and markets, the principals saw a need for a new strategic management initiative. They recognise that this stage in the company's history was crucial to its development and evolution.

SAVANT is located in an office block, named the project management centre. The workforce shares this space with two independent single project managers. They occasionally share work and expertise and regularly pool a number of administrative, clerical and technical resources. The space within the centre was only half occupied. The intention was to make this centre the administrative headquarters of the Australian Institute of Project Managers. This aspect of the company's background is of some strategic consequence and we will return to it later. The spare office space can be seen as a strategic resource; the plans for the institute reflect their attempt to 'polish the diamond' (see Chapter 2) of Australian project management and the presence in their centre of independent project managers shows their willingness to encourage strategic alliances.

The strategic management workshop

Part of the company's strategic management activity was undertaken through the mechanism of a workshop. The workshop was conducted by a leading international firm of management consultants. It followed an abbreviated version of a scheme developed by the Australian government under their national industry extension service scheme. This was designed as a national initiative to address some of the long-term competitiveness problems facing Australian industries. The strategic planning scheme embraces activities designed to move from identifying a potential

SCA to defining marketing, manufacturing, innovation, human resource and financial strategies that enable the SCA to be realised. The shorter session followed with SAVANT was tailored towards an SME in construction; the workshop concentrated on SCA identification and on marketing strategies. An SCA was defined by the workshop guidelines as:

> **"That special capability identified by the company that will enable it to attain a sustainable winning position in the market with respect to major competitors."**

The workshop was attended by ten participants. These included two directors, seven of the company's project managers and the researcher, who acted as a participant and observer. It was held over a period of twelve hours on a Saturday between 7.00 AM and 7.00 PM in the board room of the company's head office in Adelaide.

The workshop began by the management consultant facilitator asking the nine company members present to identify what they thought the company was especially good at now and how they expected to out-compete rivals in the future. This was intended to clarify any preconceived issues and the extent to which a clear and common business vision existed in advance. The workshop then proceeded to identify the range of current activities by analysing product or service groups offered and their proportion of total revenue contribution.

This was followed by a forecasting stage where product and service group growth and market share were predicted and analysis was made of prime competitors. Next came a SWOT (strengths, weaknesses, opportunities, threats) analysis from which a business mission was defined. An analysis of present and future products and markets was then made as a means to establish likely growth providers. This was converted into financial predictions. Critical success factors (CSFs) for this growth activity were isolated and their relative importance established by analysing their competitive value and the extent to which they could be used to distinguish SAVANT from its competitors. This combination of activities was then used to arrive at a defined and measurable SCA.

The workshop then moved to the key marketing issues, with particular analysis of the extent to which these issues support or create the SCA. The session concluded with a scheduling of an action plan of the key marketing and other activities to be performed by whom and by when to enable the SCA to be realised.

Profile of SAVANT activities

At the beginning of the workshop, SAVANT's current strengths were perceived as their:

- ability to win client's confidence;
- problem solving abilities;
- performance;
- ability to manage people;
- proactive management;
- desire to innovate; and
- team work.

They foresaw that this would enable them to compete through gaining industry leadership. This would arise from their:

- innovation;
- contacts;
- marketing;
- maintaining their role as an independent consultant;
- reputation and integrity;
- grasp of technology;
- systems;
- leadership;
- experience;
- unusual combination of skills;
- improving productivity;
- location; and
- background and corporate history.

They translated this into a current business vision for the future of their company which was:

- large;
- growing;
- profitable;
- a recognised leader;
- professional;
- independent;
- forward looking;
- innovative;
- specially skilled;
- able to attract the best people; and
- able to address a diverse range of problems and projects.

This was a brave statement of vision considering the deep local and national construction recession being experienced and problems with workload.

SAVANT's range of current products and markets were as set out in Table 3.2. These include a number of new product and market ventures that the company had already planned in advance of the workshop. These current and planned activities were then analysed in relation to what the overall market growth was

Table 3.2 SAVANT current products and markets

Product group	Market segment	Current annual sales (A$m)	Market share relative to largest competitor	Estimated overall market growth %pa
Pure project management	Institutions, private and government.	0.4	1:3	3
Project management support	Institutions, private and government.	0.4	1:3	8
Construction consulting	Institutions, private, contractors and government.	0.15	1:2	8
Programming	Institutions, private, contractors, other consultants and government.	0.15	1:1	8
Litigation support	Institutions, private and contractors.	0.15	1:1	–
Management consulting	Institutions and government.	0.1	Very small	>10
Future activities currently planned	Future quality systems, IT based project management, asset management, project management education, R&D.			>10

expected to be and how SAVANT's market share might change. These early strategic planning activities were aimed at identifying the growth and profitability potential of different activities. They may equate with the five forces industry analysis model put forward by Porter (1980). However, the five forces model and its approach to explaining industry segment profitability and change were never followed in the workshop. The results of the forecast portfolio are shown in Figure 3.1.

Figure 3.1 Forecast of product portfolio

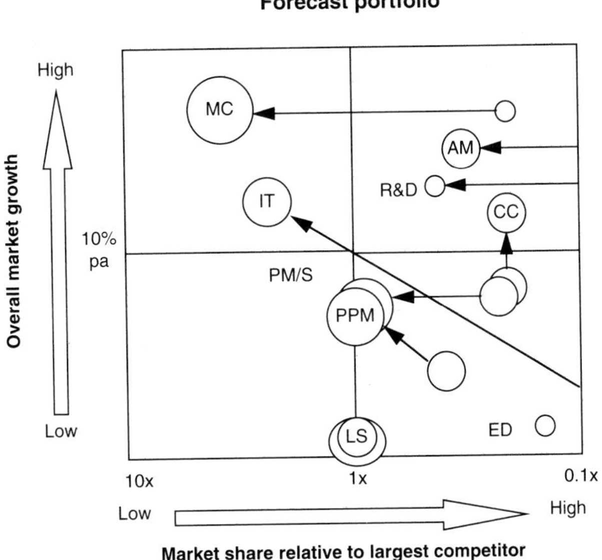

PPM, pure project management; PM/S, project management support; CC, construction consulting; LS, litigation support; MC, management consulting; IT/PM, IT based project management; AM, asset management; ED, project management education; R&D, research and development

Figure 3.1 shows that pure project management, project management support, management consulting and IT based project management were seen as the major growth providers. Asset management was also seen as a new activity with growth potential. Most others were either seen as being of low growth or as being areas where SAVANT could not obtain major market shares. The fact that construction consulting, which was felt to be one of SAVANT's key skills, is shown to have limited growth potential surprised the group and was a disappointment to them.

Analysis of competitors

This phase in the workshop was aimed at comparing the company with its key competitors. It was done in two ways. External analysis was to identify strengths and weaknesses of competitors and to identify their competitive advantages. Internal analysis was to identify SAVANT's relative strengths and weaknesses. Competitors were identified as being:

- customers doing work themselves;
- consultants and contractors;
- property agents;
- other professional project managers;
- small single professional organisations; and
- construction software consultants.

This diverse range of competitor types shows the particularly competitive nature of the consultant project management industry segment. Within each group are a number of individual companies and the analysis could have been extended to examine each major competitor type individually at the specific company level. This was the intention of the workshop guidelines. Time did not permit that on this occasion. The full analysis of these types of competitors is shown in Tables 3.3 and 3.4.

An interesting feature of this analysis is the range of competitive advantages that were perceived to be held by different types of competitor. This should convince us of the wide scope for a range of SCA types in construction enterprises.

The range of strengths and weaknesses that SAVANT is perceived to have against a range of types of competitors is noteworthy. However, the fact that some repeat over many types is also significant. From this analysis, the workshop was able to identify the key strengths, weaknesses, opportunities and threats using a traditional SWOT analysis.

The external analysis led to a perception of the key opportunities being:

- the quality of the service;
- the coming deregulation of the Australian government allocation of project management work;
- the increasing client awareness of independent project management;
- the transportability of IT systems;
- the chance of forming strategic alliances; and
- the chance of establishing a national presence.

Key threats were:

Table 3.3 External analysis of competitors

Competitor	Competitor's strengths	Competitor's weaknesses	Their competitive advantage
Customers doing work themselves	Captive market, perception of low cost, accessibility	Lack of independence, low skill level, high inertia	Political support, declared cost, incumbent supplier
Consultants and contractors	Traditional role, familiarity with client, often first appointed	Lack of independence, low skill level, poor commercial performance	Low cost
Property agents	Incumbent, constant client base, good salesmen	Vested/ conflicting interest, lack of skills, lack of focus on detail	Early contact, continuity of service
Other professional project managers	Client base, resources, geographical spread	Lack of independence, quality of service, quality of people	Geographical spread, track record
Single project managers	Price, personal service	Lack of resources, support, high client risk, narrow skill base	Price
Construction software consultants	Ability to produce software	Product support, lack of management understanding	Early start, established productive capacity

- customers and competitors exploiting their own advanced management and IT systems; and
- the lack of financial capacity and the short-term risk of excessive geographical spread.

Table 3.4 Internal analysis of competitors

Competitor	SAVANT's strengths	SAVANT's weaknesses
Customers doing work themselves	Independence, better people, innovative	Location, perceived cost, small size
Consultants and contractors	Independence, management skill	Price, timing of appointments
Property agents	Independence, management skill, industry understanding	Timing of appointment, client contacts
Other professional project managers	Skill level, small size, innovative	Geographical concentration, restricted capacity, small client base
Single project managers	Broader skill base, service security, client respect	Price
Construction software consultants	Understanding of management processes, ambition	Technical expertise, lack of a head start

From the internal analysis, the key strengths were seen as:

- the people resource;
- the interest in new ideas;
- the ability to implement project solutions;
- independence;
- reputation; and
- the combination of skills within the company.

Weaknesses were seen as:

- the lack of national coverage and client awareness;
- the poor current economic timing;
- some difficulties with internal quality and management systems; and
- the lack of resources to fund new developments.

These four components were then used to help in developing the business mission by combining the SWOT analysis shown in Table 3.5. This illustrates the four ways in which strengths, weaknesses, opportunities and threats interact. The most likely SCA (shown in italics) will arise where opportunities and strengths interrelate. For SAVANT this centred around their independence, innovation and alliances for support systems. This led to the following business mission being defined:

"We provide independent management of the total project delivery process by developing and delivering innovative solutions to asset owners nationally and to the near north."

Table 3.5 SWOT analysis

	Key opportunities Deregulation of government work, growth in independent project management, strategic alliances, emerging use of IT	**Key threats** BOT schemes, competitors' financial strengths, customers' and competitors' use of IT
Key strengths Independence, reputation for performance, innovation, resources (people)	*Most likely* Independence, innovation, supporting alliances	*Possible* Obtaining work on BOT schemes, out compete through innovation
Key weaknesses Lack of geographical spread, low client awareness, lack of financial resources	*Possible* Establish national presence, additional shareholders	*Unlikely* Build a national organisation alone, develop large individual IT and R&D programmes

The featuring of independence and innovation clearly emerges from this and the references to asset management and the near north (South East Asia) show the breadth of vision.

Identification of growth potential

The next phase of the workshop was concerned with identifying growth potential for the company. It should be remembered that despite the depressed economic climate at the time of the case study, there was a clear consensus that significant growth was desirable and attainable. This involved analysis of the present and future products and markets and prioritising the four possible combinations as shown in Table 3.6. This led to the conclusion that using present products in new markets and new products in present markets were the two areas from which growth would principally come. The new markets for present products were seen as the health, educational and defence sectors. This suggests that SAVANT were considering an emergence of a *core competence* or *core capability* in project management services that they felt could be applied to a number of industries and products. This was not developed further in the workshop or highlighted in its summary. New products embraced those shown in Table 3.2.

Table 3.6 Future products and markets

	Present products	New products
Present markets	Priority 3	Priority 2
New markets	Priority 1	Priority 4

The extent to which these two areas would lead to growth was considered by suggesting that a growth in turnover from A$1.2m to A$5m could be obtained within five years, despite short-term problems of limited work. It was anticipated that this would entail an expansion in the workforce to between 40 and 50 from the present 10.

The workshop manual definition of an SCA was as stated earlier. It is based on a special capability, enabling a winning position in respect to competitors. Much of the competitor analysis, profiling of current activities, analysis of strengths and weaknesses, and analysis of growth is intended to aid directly in identifying an SCA.

In some instances, the SCA may be obvious. In others it needs to be developed more gradually.

At this stage in this workshop, as part of a more gradual analysis, attention turned to analysing the critical success factors (CSFs) for realising the growth potential of the company. CSFs were defined as the things that must be done well to ensure the future growth of the business.

In the workshop, the CSFs shown in Table 3.7 were isolated and these were analysed. The analysis in Figure 3.2 has each of the 16 CSFs being considered in terms of the two axes. The first is whether the CSF has value as perceived by customers. The second is whether the company can differentiate itself from its competitors in this CSF. Only those in the upper right quadrant were considered to be of sufficient impact to warrant direct consideration in identifying the SCA. These are marked with an asterisk in Table 3.7.

CSFs that are excluded from the SCA by this analysis, despite their importance to early parts of the analysis, are independence, people related issues, and marketing.

Identifying a sustainable competitive advantage

The workshop had focused very directly on the importance of identifying the SCA for the company. This was seen as the prin-

Table 3.7 Critical success factors for an SCA (* = most important factors)

Critical success factors			
1	Managing people	9*	Identifying saleable solutions
2*	Technical competence	10*	Assembling a competent team
3*	Communications skills	11*	Staff motivation
4	Deadlines	12*	Innovation
5*	Mobility and responsiveness	13	Marketing
6	Flexibility	14	Disciplined internal management
7*	Ability to win confidence	15	Ability to manage and fund growth
8	Identifying opportunities	16*	Understanding client's needs and attitudes

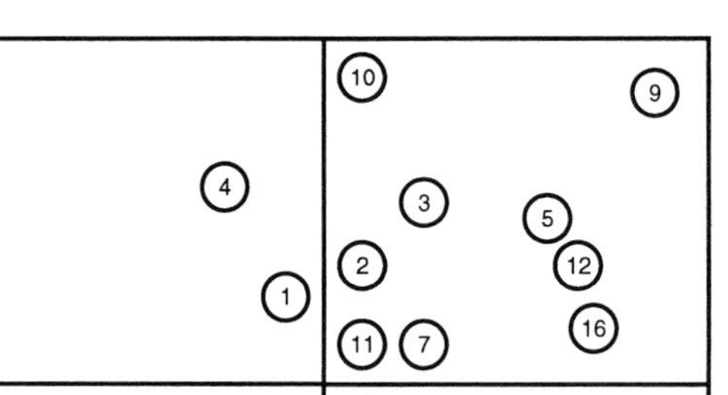

Figure 3.2 The importance of CSFs

cipal outcome and deliverable. It was defined as a key company capability which can be further developed and used as a focus for corporate development decision making. An important guideline in identifying the SCA was that it must enable the company to achieve a sustainable winning position in the market. That is, it must have at least a medium-term and preferably a long-term competitive impact. Prahalad and Hamel (1990) hold that an SCA is inappropriate in many sectors because of the increasingly dynamic context, and that a series of short term, unsustainable competitive advantages should be repeatedly sought by companies. Identifying the SCA is largely based on the foregoing analysis of competitor's growth and CSFs. The result must be something that has a pervasive cultural impact on the behaviour of all individuals in the company.

Within the workshop, the discussion in this area arose out of all of the preceding analysis. It also considered Porter's three generic strategies (1985) referred to earlier. A low-cost competitive company can be typified as a structured organisation with no-frills products that exercises tight resource controls and exploits location advantages and low assets and overheads in operating cost-driven activities. A differentiating competitive company can be typified as having strong marketing ability, creative flair, strong innovation capacity, quality leadership, with strong skill support and co-ordination. The conclusion was soon reached that the SCA would be based around product differentiation rather than low-cost service. The particular dangers of falling between these two strategies were made clear in the discussions and avoided. The workshop did not consider the second dimension of Porter's analysis in the three generic strategies. This was market segmentation. For this reason, much of the earlier attention on geographical and market diversification became lost from all subsequent parts of the analysis. In particular, the references to asset management, the near north, and the new activities in the bottom section of Table 3.2 were effectively excluded from the final SCA. The SCA may have been quite different if this had not happened.

The session therefore concentrated on a low-cost and differentiated service distinction and on competitor analysis and CSFs to realise growth. As a first stage, as was the case with much of the workshop, the large group was split into two. Each sub-group was asked to attempt to draft the SCA. The two attempts at this were eventually merged to arrive at the following statement of the company's SCA.

"We will win by understanding the client's needs and by designing innovative systems to give the client satisfaction that we provide the best solutions."

To make attainment of the SCA measurable, it was defined in terms of quantifiable criteria. Some of the alternatives considered here were:

- the extent of repeat business;
- the number of referrals;
- the cost resulting from disputes; and
- the number and value of variation orders.

Given that the SCA is expressed in terms of quality of service and client satisfaction, the two measures of **repeat business** and **number of referrals** are most directly useable as measurable criteria. It is preferable for these to be quantified as targets but this was not achieved in this instance. Benchmarking of current SAVANT and competitor achievements may have helped in this.

Marketing strategies

Having defined an SCA, the workshop then proceeded to consider how this would be implemented through marketing activities. The key marketing policies identified were to:

- monitor key clients' plans and needs and to maintain an understanding of marketing opportunities;
- monitor clients' awareness of SAVANT;
- maintain an appropriate profile and image with clients;
- monitor competition;
- maintain appropriate regional representation; and
- develop and maintain a range of appropriate products and services.

In doing this, it was anticipated that market surveys outside South Australia and construction would be necessary. It would also require that potential clients were informed that SAVANT could offer a service. This would require careful targeting of clients who match the business vision. The perception of innovation would have to be projected amongst clients. The identity and activity of key competitors would have to be better understood. This led to the scheduling of twelve key activities that would have to be undertaken to ensure attainment of the SCA and translation of it in achieving the growth and market penetration anticipated. These are shown in Table 3.8.

Summary of the workshop

The results of the workshop were the identification of a business vision and mission statement. Together with analysis of competitors and CSFs, this led to the identification of a potential SCA for the company with attainment criteria. This was translated into an action programme by considering the marketing strategies. These have been formed into a one year action plan for attaining a SCA for the company.

This description of the workshop shows many of the issues facing project management consultants and SMEs in construction generally. Of much greater value may be an evaluation of the strategic planning workshop in terms of:

- the views of its senior managers; and
- its concordance with the theoretical framework set out earlier in this chapter.

Evaluation of the Strategic Plan

Evaluation by SAVANT

The views of the two directors participating in the workshop were that, to some extent, the session had not arrived at anything that

Table 3.8 Key marketing activities to implement a sustainable competitive advantage

Activity number	Activity	Deadline (months after workshop)
1	Identify competition and review strengths, weaknesses, opportunities and threats	5
2	Set up method to gather competitive information and regularly analyse	First after 9 then quarterly
3	Identify key potential clients in Australia	5
4	Maintain potential client files and regularly update needs	Immediate and continuous
5	Conduct market awareness study for Perth and other states	1
6	Develop promotion strategy	9
7	Develop sequence and timing for interstate representation and resourcing	9
8	Select and appoint appropriate interstate personnel	12
9	Identify opportunity and entry costs for new products and tools in IT project management system and asset management	5
10	Review need for new products and tools	9 then quarterly
11	Finalise Perth business plan	2
12	Develop national business plan	12

was not known anyway. This view is supported by comparing the workshop's conclusion with some of the preconceived views held at the commencement of the workshop. This may be an indication of its success or failure – if it has clarified the right steps to take then, despite the fact that these were already known, it may be of value in itself. Much of the benefit was seen to be that all staff had participated in the event and that all were now aware of the reason for and result of the strategic plan. If an SCA is to become part of corporate culture, this type of common understanding and communication is important. It may also be much easier to attain in an SME than a larger organisation.

Although it was felt that the session brought forward few new ideas, it did make many of the priorities for the company clearer. In particular, the directors felt that the importance of identifying clients and competitors had not previously been adequately understood. The barriers to growth in certain current activities were also unknown prior to the analysis at the workshop. In many ways, the directors saw the workshop as the start of the strategic management activity of the company, rather than the end of it. They felt that the workshop had missed some of the issues and had been quick to dismiss some ideas that the directors thought would subsequently be given deeper consideration. But on the whole, the directors were satisfied.

The session represented a significant investment by the company of money and time. Much hard effort was expended over the course of a long day. Who says Australians live on beaches! An initial reaction was one of relief mixed with a hint of panic. A more considered reaction was that the workshop in itself had solved nothing. The real work in implementing and achieving an SCA remained to be done. It is important to note the directors' comment that a workshop like the one described is the start of formal strategic management activity rather than the conclusion of it.

Case study: Business process redesign by BAA

This second case study section presents a much shorter case upon which we will add further comment in later chapters. It draws upon the work of the British Airports Authority (BAA). BAA is a highly respected, experienced and progressive client of the construction industry with projects that cross most sectors. They are currently looking in detail at growing overseas markets as potential opportunities to apply their expertise in airport development and management. BAA control a large property portfolio that has to be updated and maintained and have a correspondingly high profile within the industry.

BAA reflects the concern of many clients of the construction industry. They feel that they are not getting the best possible value. This was a key assessment of the Latham review. In recognising construction clients as the driving force for the industry, Latham recommended that they should become the focal point of the industry. This accords with the theories of customer orientation of industries which are currently process-oriented. BAA have assumed a lead role in change in the construction process with a two-stage initiative.

First, the move to best practice. This is essentially a process streamlining exercise which involves the removal of process blockers to facilitate smooth process operation. This strategy is a quick-win initiative as the level of change is low and relatively easy to implement.

The second initiative is the move to world class performance. This implies a complete redesign of the project process structure which will be longer and more complex to implement. All levels of the process will be subject to redesign with the best practice of other industries playing an important part in the influence for change.

BAA see poor communications as a major blocker to the project process. They are currently producing a code of practice as a reference for all parties to the project, which sets out guidelines of information flow and responsibilities. This will be operated in conjunction with preferred partners as a virtual organisation.

Through their Chief Executive, Sir John Egan, and his recent leaderhip of a UK Government Task Force, BAA continue to be major drivers of process change in UK construction both for their own projects and for the wider UK industry. They are making detailed measurements of the benefits that their strive for lean construction is generating. They are offering major demonstrator projects as part of a Movement for Innovation in UK construction.

BAA are pursuing partnering relationships with framework suppliers from the industry and have made IT a key part of their delivery of lean construction. The latter extends to implementing IT-based process models amongst their project teams, experimenting with visual simulations of airport operations at design stage, and demanding that the industry uses IT to provide information to support their business processes. This shorter case study illustrates many of the process redesign principles being described in this book.

4 The significance of IT

Martin Betts

"A lot has changed with the arrival of e-mail. All of a sudden people want things in hours instead of within days."
– *Marketing manager, Danish dredging company*

This chapter introduces issues concerned with the strategic exploitation of IT. It shows how IT has moved into a new era in modern business by being of fundamental strategic business importance. It also introduces a number of techniques that can be used for planning for strategic use of IT. The use of these techniques is illustrated by their application in construction and other sectors. The chapter also shows the importance of business process analysis and redesign in allowing strategic use of IT.

- IT has become of strategic significance in modern business from a position of being a support technology for information processing.
- Strategic exploitation of IT requires explanation of issues of awareness, opportunities and strategic positioning.
- Awareness of how IT can be used strategically is based on raising vision through refocusing, assessing impact and defining scope.
- Opportunity frameworks for use of IT are about analysis to identify ends that are being targeted. This is done through tools for systems analysis, applications search, technology fit and business strategy.
- Positioning frameworks entail identifying means to ends in terms of scale, position and time.
- Business processes must be analysed to identify potential for redesign, major change levers and new process visions.

"The more information you have the better position you are in."
– *Business development executive, European contractor*

4.1 Introduction

We have now seen that there have been important changes in economic and business planning priorities from the short term and tactical to the long term and strategic. The new business planning priorities are no longer solely about market share and returns on investment. Increasingly, business goals are related to concepts of competitive advantage and strategic positioning.

There are a range of weapons available to business managers to attain a competitive advantage or strategically advantageous position. These include techniques of:

- human resource management;
- marketing;
- product design;
- services;
- distribution methods;
- research and development;
- use of advanced technology; and
- information and information (knowledge) management.

IT can support all of these approaches and is a key lever in obtaining competitive advantage at the corporate level. In many sectors, business managers have moved away from the historical concern with physical components of business products and processes. There is now greater concern with information-related components. IT is transforming business in terms of the nature of:

- products;
- processes;
- competition; and
- industry structure.

Technological change is a principal driver of competition that has a major role in creating new industries; IT is an equalising force between existing and new industry members but its application is widely misunderstood. There is a failure to recognise the need for technology to be used strategically. IT can be used for unification of elements within an enterprise as part of a process redesign-based business strategy, enhancing whole application areas. Business must recognise that the opportunities which IT can give go far beyond the chance to improve discrete processes, since IT has an important co-ordinating influence on business processes. The use of IT in construction has so far concentrated on discrete applications. Application of the ideas in this book is the starting point for using IT for strategic advantage, shifting from a tactical technology for internal efficiency to a strategic technology redefining the boundaries of industries. It is this, in combination with the challenging business environment, that brings the 'economic imperative of information technology' which is not yet applied extensively in the construction sector.

The emergence of the need for IT to be considered strategically has combined with changes in IT thinking. The use of IT is appropriate for many types of applications embracing:

- institutional functions;
- professional support;
- physical automation;
- external products; and
- enterprise infrastructure.

IT is now a tool for external influence as evidenced by loyalty schemes (airlines), electronic trading (banking), etc. This is in contrast to the old view that it was an internal tool for adding up the costs of doing business. IT is now being exploited to perform throughout all activities including optimisation, control and executive functions. IT is no longer a support technology managed by data processing departments but an integral part of the business manager's or senior professional's responsibility. This seems less well accepted within construction. There is a shift in the use of IT toward the strategic level and we can perceive three IT eras as shown in Table 4.1.

Surveys of both IT and business managers show information systems (IS) strategy to be commonly perceived as the most important IT-related concern. They perceive IS strategy as more important than issues of IT implementation, training, development and maintenance. Whether this is yet the case in construction is not clear. In other industries information management strategy is seen as being more important than information systems strategy.

We therefore have a situation where a fundamental change in the business environment has taken place with a move to a concern with strategic and not tactical issues. We have also seen fundamental changes in information technology. IT is now being applied to all application areas, for a variety of functions and for communications between organisations. The combination of these two forces is critical in examining the likely impact of IT. In the con-

Table 4.1 Three eras in IT (Gallagher, 1988)

Characteristic of era	IT era 1 (1960s)	2 (1970s and 1980s)	3 (Late 1980s onwards)
Objective	Transaction processing	Managerial decision support	Competitive advantage
Primary target	Clerical/ administration	Individual managers	Complete lines of business
Business justification	Productivity and efficiency	Effectiveness	Strategy

struction sector, the same forces exist although to a lesser extent. In this and the next chapter, we examine the consequences of applying these issues to the different operating levels in construction to evaluate whether we are being strategic.

4.2 The emerging IT economics

4.2.1 Different perceptions

It is clear that the information age has already dawned. We now live in an information society, where the key resource is knowledge and where IT is the key enabling mechanism. Information workers form the largest category of employees in many advanced economies and technology accounts for the highest proportion of capital formation in many sectors. Three conditions are typically required for IT to develop:

- dynamic change in technology advancement;
- commercial globalisation; and
- social advancement in education and expectation.

The IT era has replaced the data processing (DP) era and the distinction between these two is based on the much greater need for IT and its implementation to be planned and managed; IT should also be used strategically by organisations. The IT era has not necessarily dawned equally brightly in all sectors of all economies – in this and the following chapter, we examine the extent to which it has happened in the construction sector and the likelihood of it happening further in the near future.

The present technology change can be described as the IT connection. Stressing that IT has a large impact on the business of today, and that changes will have to be carefully managed, five rules have been laid down for how IT should be approached and the problems that top management will have to address.

- The key to the successful utilisation of IT is effective strategic thinking. Without an appropriate strategic perspective and robust conceptual models, it will be difficult to identify an appropriate role for IT.
- The chief strategic architect must understand the strategic nature and potential of IT and specifically manage its evolution.
- Various uses of IT may require major reorganisation at the level of work group, the department and perhaps the whole organisation.
- Applications of IT that alter the firm's core technology (closely related to its culture) may be fiercely resisted.
- Managing the IT function has become increasingly difficult as a result of these considerations.

The critical element is that it is not the application area nor its underlying technology that makes IT strategically important. It is the specific role of a particular technology application to a given industry at a point in time that makes the difference.

4.2.2 The concept of frameworks

One of a growing number of contributions to the thinking in managing IT and applying it strategically has been made by Michael Earl (1989). Amongst the issues he raises is the need for IT to be applied for strategic advantage and the need for frameworks to be used to support this. Earl describes a series of frameworks to support a range of IT management issues. Earl has labelled these frameworks awareness, opportunity and positioning. The contribution that he suggests they can make to our management of IT is set out in Table 4.2.

Each of Earl's frameworks and the models within them are a means of changing the way we look at IT. None of them in themselves offers a complete answer. However, they do contribute to a different way of thinking about technology and together repre-

Table 4.2 A framework of frameworks (Earl, 1989)

Attribute	Framework		
	Awareness	Opportunity	Positioning
Purpose	Vision	Ends	Means
Scope	Possibility	Probability	Capability
Use	Education	Analysis	Implementation

sent a powerful collection of tools to analyse strategic possibilities.

Until now, in the construction IT communities, we have not really explored these types of frameworks or the strategic issues in any depth. Do they apply to us? If so, do they apply in a different way to us than to other sectors? Who will implement and apply the models and frameworks and what impact will they have on our current and previous efforts towards using IT in the sector? Here and Chapter 5, we will attempt to address some of these issues.

4.2.3 A framework of frameworks

Awareness frameworks

This first set of models within Earl's frameworks is the most conceptual of the three types. They are intended for executives to explore the potential impact of IT and to teach themselves about areas of impact. They deal with the possibilities and the why rather than the how. They are visionary tools that are used to help change mind-sets, to suggest the scale of possible changes, and to indicate the strategic scope to a business or a sector. Earl classifies them into three categories of refocusing, impact and scoping models.

Opportunities frameworks

These tools are not intended to be visionary or educational but represent the next planning stage. They are intended to be more practical in helping analyse a sector's activities in greater detail as a prelude to implementation. The frameworks are classified into four types:

- systems analysis tools for information analysis across a business or sector;

- applications search tools which examine a business or sector for good fit with technology generally;
- technology fitting frameworks which are for specific technologies looking for problems to solve; and
- business strategy frameworks which attempt to combine IT with potential economic benefits.

These opportunity frameworks are concerned more with identifying the ends rather than the means of achieving them. They are also concerned with establishing probabilities rather than the possibilities of awareness frameworks. They are more of an analytical nature. They attempt to address the question of how.

Positioning frameworks

The first two types of models and frameworks described above are general. As we move from the general to the particular, our models and frameworks become more practical. These third-stage positioning frameworks are managerial in orientation. They relate more closely to means rather than ends. They concern implementation and capability and refer more to the question how but from a procedural rather than a technological point of view. They are the immediate preliminary to our attempt to design and implement technological solutions. The range of positioning frameworks cover scaling, spatial and temporal models. These relate to how far to take an IT implementation in terms of its context within the organisation and placing a new system in the context of other current systems and IT initiatives.

These frameworks are strategic IT planning tools that can be used to identify opportunities. Figure 4.1 illustrates the relationship between them. These relationships will be examined in detail in the next section.

4.3 Strategic information technology planning (SITP) techniques

The framework of frameworks outlined different types of model that Earl suggests can be used in SITP. We can now examine the frameworks and models shown in Figure 4.1 and consider how they may apply in construction to move us further into the IT era.

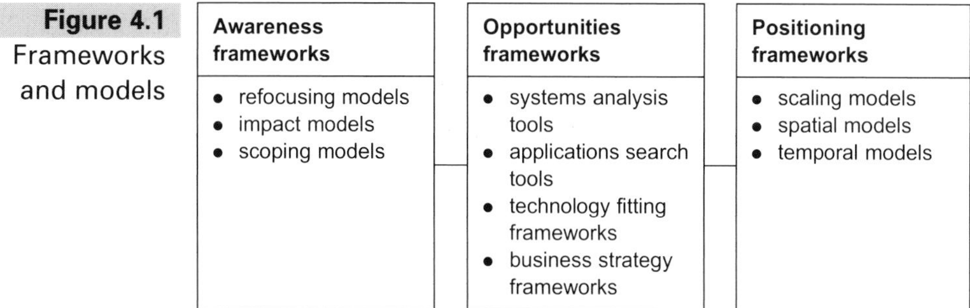

Figure 4.1 Frameworks and models

4.3.1 Awareness frameworks

Refocusing framework models

These are based on asking two fundamental questions about the use of IT. One relates to whether it can be used to significantly change current business as opposed to continuing with traditional products and processes. The second concerns whether IT should be used to change either the approach to the marketplace or internal operations. The combination of answers to these two questions are demonstrated by the matrix in Figure 4.2 called the strategic opportunities framework. Four key strategic IT systems that illustrate each of the four approaches are included in this matrix. These classic strategic IT systems are described in more detail later.

Figure 4.2 Strategic opportunities framework

	Competitive marketplace	**Internal operations**
Significant structural change	Merril Lynch	Digital Equipment
Traditional products and processes	American Hospital Supply	United Airlines

The value of the strategic opportunities framework is in raising general awareness rather than in identifying opportunities and allowing them to be implemented. Why rather than who, where or how. Earl considers this to be rather general for firm-specific use and that it needs to be tailored for different sectors. A possible application in construction is to help us think of what form of strategic IT systems we may develop. There are two ways we can do this. First by observing and classifying the type of IT systems being used by us now. Second, to identify, speculate on and specify the type of IT systems that should and could be used in the future.

Impact framework models

These models primarily address the question of who our IT systems should be developed by and for. They suggest a number of different levels at which IT can be applied. Our knowledge of construction requires us to extend these levels further as Tables 4.3 and 4.4 show.

Table 4.3 The three-level impact of IT (Parsons, 1983)

Level 1	Industry level
Level 2	Firm level
Level 3	Strategy level

Table 4.4 Five-level framework for strategic applications of IT in construction

Level 1	National construction industry
Level 2	Professional institution
Level 3	Construction enterprise
Level 4	Construction project
Level 5	Construction product

Table 4.4 shows the same five-level framework which we have referred to in Table 3.1. This time we will justify its use in the context of IT.

The globalisation of construction means that nations, and more specifically their construction industries, need to undertake strategic IT planning and devise national IT strategies. Strategic use of IT at the national level requires fundamental policy-making initiatives of a national body such as Construct IT in the UK.

The professional institutions influence many national construction industries in a very significant way. Although major deregulation appears inevitable, for now we must still take note of this professional level in IT management. This is because some countries, particularly the UK, have some of their strategic IT developments at this level. Professional level implementation is the responsibility of the institutions. Professional objectives are survival and inter-professional advantage.

In strategic IT planning, a case could be made for treating the national and professional levels as one. That may be more appropriate in identifying strategic opportunities for the future. For now, the professional level is an important and distinct one for us to consider.

The construction enterprise or company is the level that equates best with the analysis used throughout this book, much of which is drawn from industries where other non-enterprise levels are not so relevant. The most significant and common examples of current strategic IT planning in construction occur at this level (Betts, 1992). They are usually from major international contracting firms but also include component and materials suppliers and consultants. The vast majority of construction enterprises, of a design or construction nature, are very small. The extent to which strategies can embrace both ends of this spectrum is a problematic issue. Future implementation at the enterprise level may be of most commercial significance in the short term but we must recognise the variability that exists between enterprise size. Many innovative strategic initiatives for using IT will emerge in the immediate future if business development and strategic management teams become widespread. The consequences of this will be restructuring, the creation of new businesses and the emergence of new information intensive products and the offering of improved services to clients.

In contrast to business and manufacturing industries, the fundamental operating level of construction is the project, regardless of size of enterprise. This level equates with the industry process. It explains our current pre-occupation in construction IT research with project or process models. At the project level, implementation and consequences will differ again. Regular developers and clients of construction will have opportunities to exploit strategic IT planning. New client demands based around IT are conceivable. There will be software houses and project information management consultants with potential to exploit competitive advantage through IT.

These different levels have been used in a national strategic IT plan for construction in Singapore. As a contribution to the extensive national infrastructural planning efforts, a communications network, in the form of an impact model, was drawn up for presentation to and discussion among local industry members. The framework has been of some consequence within Singapore in initiating further frameworks and developments by other parties who have been made aware of the impact that IT can bring to the sector. The principles of the framework are depicted in Figure 4.3. As an extension of the research project of which this framework was part, each of the major professions and types of enterprises have been examined for their current activities and these have been classified as areas where the impact model could support the current processes as in Table 4.5.

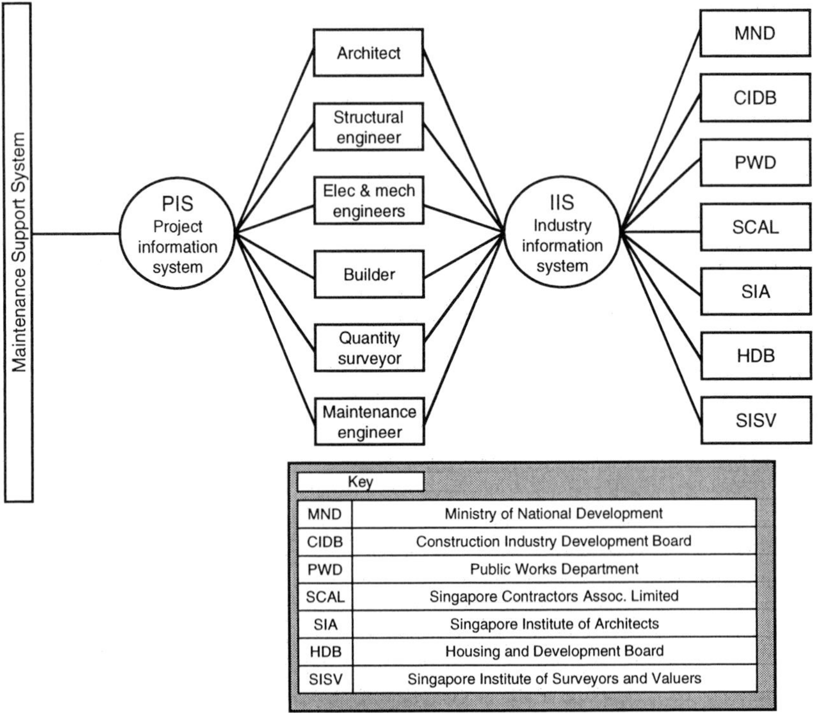

Figure 4.3 An impact model for construction in Singapore

Scoping models

These are again a means of exploring general awareness of strategic potential within a sector by analysing the information component of a sector's processes and products. The suggestion is that there is little scope for exploitation of IT in a sector or part of a sector where the information content is low. Where are the high information components of the construction sector? Figure 4.4 attempts to address this question by analysing which aspects of construction have the greatest information content in their process

Table 4.5 Analysis of industry information systems requirements

Description	Government information system	Professional information system
Consultant appointment		✓
Recruitment of new staff		✓
Preparation of specifications		✓
Nomination of contractors for tendering	✓	
Final accounts for completed projects	✓	
Maintain records of material prices	✓	
Results of price enquiries to suppliers		✓
Compile building tender price index	✓	✓
Compile building materials cost index	✓	✓
Application for permits before site commencement	✓	✓

or product. The number of new computer systems and current research initiatives that seem to address doors and windows because they fit demonstrations of the technology rather than the need and scope suggests this form of model may have application in providing us with a clearer idea of the strategic scope.

A model like this one identifies where our first steps towards more effective use of IT in construction should be made, if we are to be working with the strategic business forces rather than against them. It shows where the scope in construction is. These areas are identified by Figure 4.4 as being in architectural and engineering design, software development, database creation and distribution. These areas are characterised by high information intensity in both the process and product. They are not the areas where we usually concentrate our IT implementations. They may appear peripheral to the mainstream activity of much of construction. However, the model suggests that this is where we should concentrate our

Figure 4.4 Information intensity matrix (after Porter and Millar, 1985; Hannus, 1992)

		Information content of product	
		Low	*High*
	Low	Delivery and assembly of low-tech building sub-systems, e.g. windows, doors, concrete/steel beams, excavation, in-situ concrete casting	Delivery and assembly of high-tech building sub-systems, e.g. elevator, access control system, remote maintenance monitoring system
Information content of process	*High*	Construction process in planning, estimating, managing and control stages	Architectural and engineering design. Software development. Database creation and distribution

implementation and that we should avoid those areas that the scoping model is drawing our attention away from. The depth and sophistication of the analysis shown in Figure 4.4, and in many of the other figures and tables presented in this chapter, is insufficient for the purpose of making a final commitment on such issues. What this and other figures do is to allow us to make a start in the strategic planning analysis required within our sector.

4.3.2 Opportunities frameworks

Systems analysis tools

There are three basic approaches behind these tools. The strategic approach is typified by the value chain. The information flow approach is represented by data flow diagrams. The technology approach is typified by process and product models.

A value chain example of a systems analysis framework was shown in Figure 2.2. As we saw in Chapter 2, value chains can be used to identify potential for competitive advantage within individual parts of the whole firm. The value chain is a structured way of analysing a business's constituents and its links to outside organisations. Value is defined as what a company creates, measured by the amount buyers are willing to pay for the augmented product or service. The difference between value and cost determines profitability. Value chains can be used to identify lower cost, higher value and value channel linkage applications. The parts of the chain include an organisation's infrastructure as well as different categories of its direct productive processes.

The use of data flow diagrams is one of a number of means by which information flow within processes can be traced. The application of this technique to parts of the quantity surveying enterprise's activities in Singapore is shown in Figure 4.5. This was drawn in support of the impact model in Figure 4.3 and Table 4.4. The study that was made was to trace information flow within the primary enterprise types in the Singapore construction industry. It was based on extensive, iterative, site-based studies within a sample of organisations. This was done for architects, contractors and quantity surveying enterprises. Others have looked at the information flow process using other techniques. These information flow tools help home in on opportunities.

The rapidly growing array of process analysis techniques might also be described as systems analysis techniques. These would include process mapping techniques and process modelling based on conventions such as IDEF0, a technique we will cover again in Chapter 12 which examines benchmarks of project information integration.

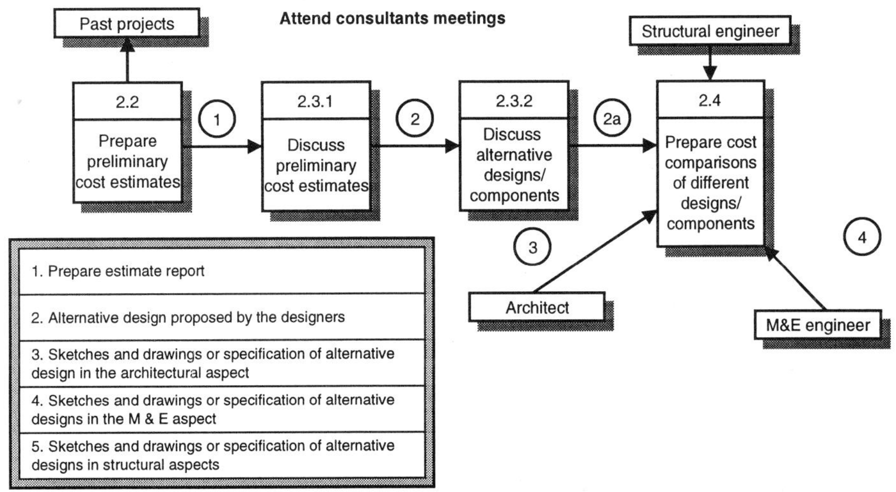

Figure 4.5 Example of a data flow diagram

Applications search tools

These help identify where IT applications should be made within an economic process. An example is the life cycle checklist proposed by Ives and Learmonth (1984), shown in Table 4.6.

This technique views the life cycle of an economic process from the customer's viewpoint rather than the producer's. It will usually

Table 4.6 Customer resource life cycle (Ives and Learmonth, 1984)

Life cycle stage	Typical question	Construction example
Requirements		
Establish requirements	How much of the resource is required?	Briefing and feasibility
Specify requirements	What are the required resource's particular attributes?	Design specification
Acquire		
Select source	From whom will the customer obtain the resource?	Procurement model
Order	How will the customer order the product?	Selection
Authorise	How will the customer pay for the product?	Tendering
Acquire	How, where and when will the customer take possession of the resource?	Negotiation
Test and accept	How does the customer ensure the resource conforms to specifications?	Alternative design proposals, design competitions
Stewardship		
Integrate	How is the resource merged with inventory?	Commissioning
Monitor	In what ways can the customer monitor the resource?	Use, property management
Upgrade	How will the resource be enhanced if conditions change?	Facilities management
Maintain	How will the resource be repaired if it becomes necessary?	Refurbishment maintenance
Retirement		
Dispose of	How will the customer move, return, sell or dispose of the resource when it is no longer required?	Demolition, disposal, redevelopment
Account for	How much is the customer spending on the resource?	Life cycle costing

entail a broader view than that normally taken by organisations, particularly within highly fragmented sectors like construction. Many IT strategists suggest that systems beyond current boundaries of organisations offer strategic opportunities. Our first reaction to this table should be to ask ourselves where we are doing all our innovation and making all our IT applications. The answer appears to be mostly within individual parts of the life cycle relating to single organisations. Why is this the case? The answer appears to be the fragmented nature of the industry and IT may change the nature of this obstacle. In many ways the message of this model is that an effective overall construction process requires IT. This is commonly experienced in other sectors, where inter-organisational, whole life-cycle information systems have delivered many benefits.

Technology fitting frameworks

These forms of framework require that we examine current systems and emerging technology to identify opportunities for technological progression. They should be based on evolutionary rather than revolutionary system development. The research community seems to be focused on looking to fashionable technologies like expert systems, neural networks, virtual reality and object-oriented systems and trying to find problems that the technology can solve. This is not necessarily the wrong approach but we must recognise that it is only one possibility. The way that management and business and technology are coming together in many spheres of economic activity has highlighted the difference between technology-push and strategy-pull as the principal drivers of technological innovation. This form of framework is based on the technology-push driver. Many of the others we consider here are strategy-pull.

The analysis in Table 4.7 shows the way the activities of the architectural profession and its enterprises in Singapore were analysed to identify possible applications of expert system technology. These resulted from analysis of the data flow diagrams, an example of which is shown in Figure 4.5. The data flow diagrams were themselves in support of the impact model in Figure 4.4, which shows how models and frameworks at these different levels and stages interrelate.

Business strategy frameworks

Many of these models are those considered in Chapter 2 in our review of strategic management techniques. The five forces model

Table 4.7 Areas of expert system application to architects

Process	PIS	EIS	IISG	IISP
Draft scope of work	✓			
Decide on professional fees		✓		
Conduct feasibility study based on market data	✓			✓
Produce spatial requirement and brief	✓			
Amend spatial requirement and brief	✓			
Propose various designs	✓			
Decide on form of procurement	✓			
Select contractor	✓			

PIS Project information system
EIS Enterprise information system
IISG Industry information system (government)
IISP Industry information system (professional)

can be used as a means of positioning an enterprise in relation to market forces particularly through exploiting industry changes. The five forces within the model are:

- buyer power;
- supplier power;
- threat of new entrants;
- substitutes; and
- jockeying for position.

The model can be used to identify where IT can be used to prevent new competitor entrants, to exploit buyer or supplier power relationships, to offer substitutes, and to help in jockeying for position between competitors. The combination of business forces in Figure 2.1 shows the nature of the competition within an industry segment. This is seen as a key framework by which to explore potential IT applications. Table 4.8 illustrates Earl's (1989) interpretation of how IT can apply to these five forces and some examples we have added from construction.

The construction examples included in this table must be

Table 4.8 Exploiting IT in the competitive arena (Earl, 1989)

Competitive force	Potential of IT	Mechanism	Construction example
New entrants	Barriers to entry	Erect/ demolish	BCIS and ELSIE (*professional*)
Suppliers	Reduce power	Erode share	COMIS Owens-Corning Fiberglass Corporation (*national/enterprise*)
Customers	Lock in	Switching costs Customer information	Steetly Building Products (*product*)
Substitute products and services	Innovation	New products Add value	Owens-Corning Fiberglass Corporation (*enterprise*)
Rivalry	Change the basis	Compete/ collaborate	CLIENT real-time project information management (*project*)

described. BCIS (an industry wide professional information system) and ELSIE (an innovative expert system product) are reflective of an approach in the UK where strategic management by individual professional bodies is common. At this level, shared information, and IT systems development, for subscribing members of professional bodies is for their individual benefit. These examples come from the quantity surveying profession. The first project is an on-line data sharing service of elemental cost analyses for design stage cost estimating. It is open to all profession members. This is an example of an inter-organisational information system. Its aim is to improve the design stage building price estimating service of the profession in general. A second initiative was participation by the QS profession in the Alvey research programme. This was a UK fifth generation computing research initiative. The project was to build an expert system to aid in early project decision making. Access to the research product was originally denied to non-members of the profession in an attempt to erect a barrier to new entrants offering quantity surveying services. There is little evidence to date of the profession exploiting technology for links between their members and others in the industry.

The Construction Industry Management Information System (COMIS) was a development in Singapore jointly undertaken by the Singapore Contractors Association Limited (SCAL), the National Computer Board, the Construction Industry Development Board, and various other national public bodies. It was a generic specification of a management information system for Singapore contractors. Its likely impact on suppliers, given that it could be developed for all local small and medium sized enterprises, would be to reduce supplier power.

Owens-Corning Fiberglass Corporation is a US home insulation enterprise that allowed builders and designers free use of energy efficiency evaluation software in exchange for guaranteed incorporation of insulation products. This is an old example of a service that was made possible by IT tied to the continued use of what was an existing, highly competitive product. It exploited supplier power as well as preventing substitute products and services penetration of the market.

Steetly Building Products was an independent brick manufacturer in the UK that was subsequently taken over by Redland. Their Parkhouse Brickworks produced 1.6 million bricks each week in a fully automated plant in the early 1990s. The order handling system they developed was able to increase flexibility in the scheduling of production. The result was a large reduction in inventory build-up and order backlog. The system also allowed architects to connect their CAD systems directly to the plant in evaluating the influence of design on brick requirements (Bradshaw, 1990). A connection between the customer's and producer's systems produced accurate, volume-based ordering. This was a highly strategic development that locked in potential customers in a similar way to the commonly cited American Hospital Supply (AHS) company example (see p. 109). It is an application of strategy for what may be thought to be an unlikely strategic target and demonstrates the breadth of possibilities in our sector.

CLIENT is a commercial product, by a software supplier. It aims to provide total project information management and acts as a central monitor of the status of information production and circulation between all members of the project team (Hemmett, 1991). As many as eight individual organisations including architect, engineer, QS, builder and subcontractors use a dedicated terminal and modem link. All project communications at design and construction stage are through the system and monitored by the owner or project manager. The owner pays for the terminals, modems and software used by all the participants, treating it as a project overhead. The system is marketed not to architects, engineers and builders but to clients. Clients choosing to use the system are stipulating compulsory use of the software to all other organisations when discussing engagement terms. It is being used on several major projects in Australia and East Asia. It is a system that exploits value chains and buyer power as well as being influenced by and influencing the rivalry and nature of jockeying for position between industry members.

4.3.3 Positioning frameworks

Scaling models

A scaling framework refers to how far to take IT systems in terms of their strategic importance to an organisation. Are they critical to the business activity of the organisation or more simply systems that aid the business process? This can be addressed by again answering two questions:

(1) How quickly would failure in an IT system come to the chief executive's attention?
(2) Is IT development one of the five things that must be right for the future survival of the organisation?

An example is a major American financial institution where failure in their IT systems was immediately notified to their CEO on the Friday afternoon when it happened. Business failure may have resulted if the system had not been reinstated soon after business resumed on Monday morning. This system was clearly of high strategic impact.

In the grid of Figure 4.6, some examples of generic and proprietary types of IT systems in construction have been classified. Accounting systems were one of our first and remain the most common construction IT application. Their strategic impact is limited and might remain so. CAD as currently used is of limited strategic impact at present although important as a productivity aid. However, it is likely to be of key importance to the future of the design professions. Bills of quantities systems, as used by quantity surveyors and contractors in different parts of the world, are unlikely to be of key strategic importance in the future as professional demarcations deregulate and as data exchange becomes more sophisticated and as business processes change. However, in some cases at present, IT systems are of critical importance. Manual systems have been extensively imitated and automated by IT systems in some enterprises. In the rush to prepare a key tender, failure of one of these systems at the time that the bid is to be submitted could be of strategic significance.

Figure 4.6 Position of information systems: construction examples (source: McFarlan, 1984)

Strategic impact at present	Strategic impact in the future: *Low*	Strategic impact in the future: *High*
Low	**Support** Accounting systems, office automation, word processing (*all levels*)	**Turnaround** CAD (*professional*)
High	**Factory** Bills of quantities systems (*enterprise*)	**Strategic** OTIS elevator maintenance management (*product*)

Finally, the OTIS elevator maintenance management system. OTIS used IT to replace its decentralised service system in the US in the 1980s. Elevator service contracts are a highly profitable aspect of building facilities management and to strategically ensure a valuable and sizeable share of the market, a system called OTISLINE was developed. It does not use particularly sophisticated computing technology but is a very strategic IT application. The system provides weekly service reports from a centralised database. This helps to isolate trouble spots and improves the speed and quality of service calls in a way that increased the value of the service to building owners and led to OTIS being able to offer a differentiated service. Under the old system, field officers, who were hard pressed dealing with service calls, would only provide summary reports from which problems could not be identified and which prevented the quality and speed of service from being improved. The design, assembly and installation of elevators is not a highly profitable service and offers little strategic opportunity but it may be an area where our technological advancements would encourage us to make IT applications. Elevator maintenance has been identified by OTIS as a key strategic activity and the IT system they have designed in a 'strategy-pull' rather than 'technology-push' form of innovation has been a key factor in their business success. The example should provide inspiration to us as potentially offering further opportunities in our continuing construction IT innovation.

Spatial models

Spatial frameworks are appropriate for analysing whole industry sectors and their general characteristics. Earl (1989) has commented on assessment of IT impact with metaphors for the strategic context of IT in different sectors. Table 4.9 shows this classification together with some examples from construction that we can apply to the metaphors.

BCIS is a system we have already detailed (p. 95). The Royal Institution of British Architects (RIBA) offered a similar form of use of IT through the RIBACAD system (Ray-Jones, 1990) library of standard architectural details. In both these examples, the complete delivery of parts of professional services is through an IT system. We have also already described Steetly Building Products (p. 96). It is relevant because the product's future is increasingly dependent upon the success of the IT systems, in contrast with other firms in the market. CAD data exchange is an area where many construction commentators have speculated on emerging strategic opportunities. There is evidence of some of the integrated architecture, engineering, construction (AEC) contractors beginning to exploit them. Further opportunities exist here at the national and professional levels. Finally assembly work on building sites is a delayed activity. Robotics are often examined as strategic opportunities but these appear delayed at present at the implementation level.

Temporal models

Although innovation sometimes presumes this to be the case, effective use of IT in construction is not achieved in one step – many of the disappointing early experiences with IT applications in practice failed to take account of this. The achievement of improved IT support to effective construction processes must be managed and gradually achieved. These frameworks address this issue and relate to time and the level of managerial and techno-

Table 4.9 Sector framework for IT (source: Earl, 1989)

Strategic context	Characteristic	Metaphor	Construction example
It is the means of delivering goods and services in the sector	Computer-based transaction systems underpin business operations	Delivery	BCIS, RIBACAD (*professional*)
Business strategies increasingly depend on IT for their implementation	Business and functional strategies require a major automation, information or communications capability and are made possible by these technologies	Dependent	Steetly Building Products (*product*)
IT potentially provides new strategic opportunities	Specific applications or technologies are exploited for developing business and changing ways of managing	Drive	CAD data exchange between design team members with industry reorganisation (*national/professional*)
IT has no strategic impact in the sector	Opportunities or threats from IT are not yet apparent or perceived	Delayed	Assembly work on building sites (*project*)

logical progression. Their importance and contribution are to help us to understand the nature of our gradual progression in the use of IT. The construction examples are classified in this way to help us appreciate:

- the nature of innovation with different technologies;
- the stage and nature of implementation of different technologies; and
- why they are at different stages.

One can also observe competitive progression in IT developments in a sector. Our understanding of how national level IT applications are progressing over time can be shown as in Figure 4.7.

Table 4.10 Technology assimilation and management (source: McKenney and McFarlan, 1982)

Stage/factor	Technology identification and investment	Technological learning and adaptation	Rationalisation/ management control	Maturity/ widespread technology transfer
Challenge	Identify technology of potential interest and fund a pilot	Encourage user experimentation on broader base	Develop tools and techniques for efficient use of technology	Adaptation to and of technology
Goal	Technical expertise and early application	User insight on potential User awareness of technology	Value for money, reliability and longevity	Diffusion Integration
Management	Lax planning and control	Encouragement and observation	Standards, analyses and studies	Organisational processes
Growth processes	Technological advance Application testing	Applications advance User learning	User advance Management learning	Management advance
Construction example	Neural networks for planning and estimating	Expert systems	CAD	Project management software

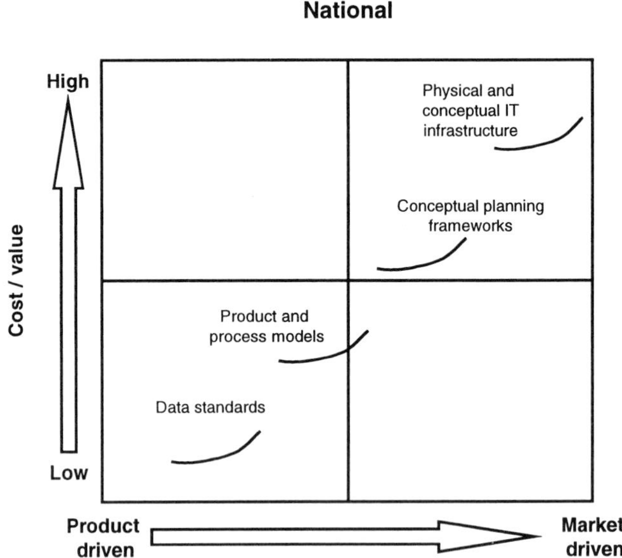

Figure 4.7 Competitive progression in national level construction IT applications

4.4 Business process analysis techniques

Having described Earl's frameworks for strategic IT planning we can now describe how business process re-engineering may be undertaken with IT. This is a further technique in planning that is strategic in its nature and has IT as its focus as we have already seen. It is a methodology that can be combined in use with some of the other techniques described in previous sections of this chapter. We propose a methodology for applying BPR to organisations using the following five key phases:

- identifying processes for redesign;
- identifying change levers;
- developing process visions;
- understanding existing processes; and
- designing and prototyping new processes.

4.4.1 Identifying processes for redesign

This phase must follow a strategic planning exercise to determine the aims and objectives of the company. The objective of BPR is to design a process structure for the company that procures the strategic plan and moves the company effectively in the right direction. The process structure will probably look very different from the existing structure of the company to reflect the cross-functional nature of processes. As mentioned earlier, there is an optimum number of processes that a company should target as a sensible structure and core processes are of particular importance. The concept of discontinuous thinking is essential during this phase to ensure that processes are not old function-based solutions to strategic planning but 'clean sheet' approaches that forget the past.

4.4.2 Identifying change levers

For BPR to be effective within any company, the organisational infrastructure must change to facilitate the process structure. Change levers come under three broad headings:

- information;
- IT; and
- human resources and structural change.

Information and IT

It is important here to disentangle information from IT. Information is 'the words, numbers, images and voices that impart meaning and inform', while IT is the technological basis for the capture, manipulation, storage and communication of information. Studies have estimated that over 85% of the information within most companies is not on a computer system and that managers get two-thirds of this information from face-to-face or telephone conversations; the remaining third comes from documents from outside the organisation which are also not on the computer system, although this situation is changing rapidly. People in organisations do not share information easily because it maintains their power; it is precisely this inability to share information that inhibits effective process operation. It is a major barrier to many organisations currently pursuing knowledge management strategies in all industry sectors. Further problems are that there can be fundamental differences about the meaning of even the most basic item of information and that information should be filtered as much of it is irrelevant. This is an increasing concern with the information overload that voice-mail, e-mail, fax and mobile phones have created.

Information architecture has attempted to structure information in organisations and this is of considerable use to BPR. First, there should be a shift from unstructured to structured information oriented processes, Second, information requirements should be defined from capture, storage, manipulation and communication. Third, IT is an enabler and not the focus for information management. These ideas attempt to add tangibility to information, when most information is intangible. To improve the information flow requires changing basic behaviours, attitudes and incentives that relate to information.

Human resources and structural change

Probably the most difficult factor to change in a BPR initiative is the behavioural characteristics of employees during a period of great upheaval and change. Machiavelli observed the unrelenting human ability to resist change, but employee support for BPR can in fact be a most positive lever to enable successful change. Unfortunately managers often find employees putting up every barrier within their capabilities to maintain the comfortable status quo.

BPR recognises the importance of the human element in organisations. A new process structure with corresponding information architecture and IT will not have the desired impact on productivity unless employees use it in the desired way. This means changing the way employees behave and work.

Cross-functional processes are mirrored by cross-functional teams. Numerous studies have proved the benefits of organising work units into teams to improve productivity. Worker isolation was a principle of Taylorism, under the assumption that the more isolated workers are, the harder they will work. Cross-functional teams have the benefit of including the expertise of several functions within a single work unit. Many companies now use cross-functional teams for new product development so a manufacturing representative can advise about the practicalities of building the product and marketing and sales representatives can decide whether the customer actually wants the product, and how easy it will be to sell it.

Within a team structure, work life tends to improve since people crave social interaction from employment. When the team is tackling informational work, social interaction facilitates the passing and sharing of information and problems so team synergy can be used to deal with a problem instead of the lone efforts of an individual. Of course, teams can have a negative effect when functional diversity causes conflict and impedes innovation.

The need for multi-functional skills requires individuals who can deal with a variety of work. As the content of work becomes more substantial, it also becomes more rewarding as the workers' knowledge and understanding of the overall process and their contribution become apparent.

Hierarchical structures within a process environment 'flatten' to allow businesses to become more responsive and flexible to external pressures. If information has to pass through endless levels of hierarchy for any decision to be made, then a company will never react quickly to competition. Layers of middle management are stripped away and employees are empowered to make their own decisions rather than delaying workflows for several managers to consult. Empowered workers are more productive as they have more power to control their work life which has a positive motivational effect.

Advancement and remuneration policies change in a process environment as the criteria change and the hierarchy flattens. It is not assumed that competence at one job should guarantee promotion to another. Good bricklayers do not necessarily make good foremen. Promotion to foreman may mean losing a good bricklayer and gaining a bad foreman. Workers will be paid based on their performance. Taking an extreme view, if a bricklayer's productivity is outstanding and passes every quality inspection then he should be paid more than his supervisor whose performance may be mediocre at best.

Just as BPR reassesses business operations it also reassesses the way we work and whether our traditional values for working are the best to ensure optimal organisational performance.

4.4.3 Developing process visions

This phase is complementary to the selection of processes for redesign. In the development of process visions, the operation of the new process structure becomes more focused and objectives for performance are set.

As BPR teaches us that attention to business networks is essential, there should be consultation with key clients to gauge their expectations and needs in interfacing under strategic alliances. Stakeholder analysis is an important tool in assessing who has a vested interest in the output of the process. Even though the process may not cater for all stakeholders it is nevertheless important to appreciate how the process will affect them.

Benchmarking is a useful tool to assess how competitors or different industries manage their processes. Benchmarking was discussed briefly in Chapter 2 and is described more fully in later chapters.

Process metrics are an essential measurement tool in setting objectives for performance. Rather than setting performance targets like improving quality in the construction phase, which is ambiguous and unclear, objectives of cost, performance and quality should be measurable. Although this may initially seem uncertain for construction due to the uniqueness of each project, averages for performance targets can be set from past experience. For example, it should be relatively easy to compare project value and time to prepare tender price to measure performance. Armed with this information, improvement metrics can be set. It may be the desire of the company to reduce tender preparation time by 50%. This particular target may seem unreasonable, but with effective team working of planning, estimating and purchasing professionals and appropriate technologies, serious improvement metrics can be set. Even if only 40% of the target is achieved this is still significant. Such defined and measurable performance improvement targets are at the heart of the Egan Task Force Report (1998) and its Movement for Innovation in the UK.

4.4.4 Understanding existing processes

In exploring existing processes, the BPR team is not looking for in-depth knowledge and process mapping of the current process structure, but a high level understanding of how they work. This exercise should help the BPR team to appreciate what is wrong with current processes and why improvement is essential. Understanding why a process has been inefficient is instrumental to ensuring that the same mistakes are not repeated.

4.4.5 Designing and prototyping new processes

The design and prototyping of new processes is the main coordinative activity of process redesign. Detailed process maps are drawn based on a fundamental knowledge of how the process should operate. Some have suggested two approaches to designing and prototyping. First is the high impact approach where mission critical processes are addressed and redesigned. These processes are expected to have maximum business impact. The second approach is exhaustive, where all processes are redesigned. They observe that most companies cannot afford the time or to deploy the resources using the exhaustive approach, so the high impact approach may be more pragmatic in many situations.

The prototyping of new processes might be performed by a pilot study method. In a large construction company that operates with many regional offices, it may be advisable to test the process out in one office before implementing it elsewhere. Despite detailed process mapping, the BPR team will often fail to predict operational side-effects. Human resistance to change can be a powerful force if not managed effectively, especially where radical change is involved.

4.5 Summary

The aim of this chapter has been to examine how the range of frameworks and models used by IT strategists in other sectors also apply to construction. The speculation was that they may provide a strategic contribution to our current technological progress towards innovation in construction IT. The aim of this chapter was to be broad and identify scope rather than to be deep in analysis and in confirming opportunities. Much more analysis remains to be done and we must address a further issue which is a problem of methodology. This concerns the question of whether our analysis of strategic opportunities should be based on our observation of what is happening or speculation and specification of what could and should happen. In the latter case, the problem is who should speculate and specify implementations and under what authority they do so. This should be economic and business authority rather than theoretical and technical authority. We are also constrained by the issue of whether our strategic speculations should be based on our current industry organisation or our assumptions of how the industry will be organised in the future. As we have seen, the fragmented current picture is a strategic barrier. This is the main extra dimension that the section on business process redesign seeks to add.

The main results of the analysis in this chapter are that many of the models and frameworks do apply. We can classify much of our current practice in construction by these models. The models cause us to sit up and think about what we are doing and the practical applications we are making. They particularly cause us to ask the questions where, why and how far to go with IT applications. These questions have not been asked extensively before now. This analysis shows the need for construction to be different from other sectors and primarily answer the question who should apply IT strategically.

There are more possible answers to this for us and the five levels we have identified each have quite different requirements. We have attempted to show the requirements of all five in this chapter but in doing so we are only scratching the surface of the problem. The analysis of all the models within each of the three types of frameworks could be made separately at each of the five levels. This gives rise to a problem we will continue to face. The overwhelming scale of opportunities and possibilities may prevent us focusing on the best. The achievement of more effective IT support to improved construction processes may be technologically possible but to be a business proposition the implementation issue must be rationalised. A conclusion that we can draw from our analysis is that the construction examples we have used are exceptional. The general picture is that much of our current construction IT practice is immature compared to other sectors and to the theory. The implications of these conclusions are that our construction IT applications may currently be misguided in being too technologically oriented.

We can draw an analogy here with the rainmaker theories adapted from Armstrong (1985).

> Stage One: "There is a drought! Let us appoint a rain dancer."
> Stage Two: "The dancer is dancing but it is still not raining."
> Stage Three: "Yes I know it didn't rain but didn't you like the dance?"
> Stage Four: "Who cares why it rains? – I'm a dancer."

Are researchers and construction IT application managers forgetting why they are advancing technologies and the purposes to which they will be put and the extent to which they will be used?

Finally in this chapter, we briefly discussed a methodology for BPR. There are a variety of tools that can be considered. As we will see in more detail later, these include process mapping techniques and soft systems methodologies. These are further explained in Part C of this book, which is concerned with the way forward. The methodologies require iteration. Despite being portrayed as linear, BPR entails revisiting stages of the methodology. Davenport (1993) does not advocate a strict ordering of the stages; the only stipulation he advises is to identify processes for redesign before understanding existing processes, as discussed above. We will now consider a number of case studies, which illustrate some of the points raised in this chapter.

Case studies: Non-construction strategic IT planning

These three examples illustrate the way IT is used strategically in some other sectors. A short description of some of these examples, taken from Earl (1989), will help illustrate the principles discussed in this chapter and their potential for construction.

American Hospital Supply (AHS), the US medical supply company, provided on-line order entry terminals for purchasing executives in hospitals. AHS carried a very broad product range and hospital buyers could use the terminals to enquire on stock availability, price and delivery and then order. So attracted were purchasing executives by the system that acquisition of an AHS terminal was very acceptable and orders increasingly were placed with AHS. Ordering and distribution costs were reduced for both AHS and their customers. In a market of over 400 rivals and 7000 major customers, AHS achieved 17% per annum growth in market share in the late 1970s and early 1980s. The system changed the traditional relationship between customer and supplier.

Digital Equipment Corporation (DEC) built an expert system, XCON, to develop computer configurations for customers, building on the accumulated knowledge of their design and field service engineers. This reduced costs of rework, reduced installation delays and thus improved both DEC's cash flow and customer satisfaction. The system is also used by sales personnel for specifying alternative configurations in the order generation process. XCON thus represents a radically new approach to internal logistics and their management.

United Airlines adopted teleconferencing for operations management in emergency situations and daily executive briefings. It allowed efficient and effective communication between key personnel located in dispersed airports and controllers at headquarters. Teleconferencing has been used subsequently in labour negotiations across the airline's dispersed network. These are applications of new technology to traditional processes.

Case studies: Non-construction applications of business process analysis

Some of the following examples and others can be studied in more detail in the work of Hammer and Champy (1993).

The Ford Motors Accounts Payable System

The Ford Motor Company in North America, exposed to increasingly dynamic market and competitive conditions in the late 1980s, sought to rationalise its administrative systems. Up to that time it had operated a suppliers and parts payments system based on paying on demand. The issuing of invoices, delivery notes and orders were each handled separately and all had to be checked individually against each other. The large number of instances when they did not tally had to be individually investigated. This system was substantially rationalised under a major process-based change project. The new system entailed a buyer entering orders onto an on-line database. If this order was fulfilled by a delivery, it would be automatically accepted and paid. If the delivery did not match the order, the goods would not be accepted. Invoices were completely removed from the process and the numbers of staff engaged in processing transactions fell from 500 to 125.

The systems had effectively moved from payment-on-demand to pay-when-delivered. Its effects on process (beyond time and cost improvements internal to Ford's administrative systems) are that it encourages avoidance of early delivery, thus reducing double-handling and storage space requirements, and gives an important stimulus to just-in-time (JIT) logistics management.

Kodak Concurrent Engineering

This process redesign example arose from the time-based competitive advantage gained by a major Kodak competitor. Fuji's launch of a single-use camera had a significant market impact when first launched and caused Kodak to think quickly and seriously about how they could respond. They had a pressing commercial need for high speed product development to remove the unique competitive advantage that Fuji was temporarily holding.

The response was to explore principles of concurrent engineering, facilitated through the use of a shared database with which remote design teams could interact. This system of work was a departure from their traditional patterns and entailed parallel, independent design from discrete engineering specialists but with dynamic status-checking of the work of others through the shared database.

Hallmark Cards Simultaneous Engineering

This case also related to commercial pressures to speed up design cycles. The solution in this case might be better described as simultaneous engineering in that it was more closely based on team working by physical sharing of space between design specialists and process changes.

Wal-Mart Supply Management System

This involved a collaborative supplier arrangement that developed between a retail outlet and a major supplier of baby nappies. The retailer established that the sporadic replenishment of high stocks of low-value goods was proving expensive and appeared to be poorly managed. They eventually outsourced their inventory management activities to their suppliers by developing a close working relationship through which the supplier was granted access to on-line inventory depletion data and could arrange delivery schedules on a continuous replenishment basis for the retailer.

Corning

Corning is a technology company situated beside the Appalachian Mountains in the USA. It supports the local community, employing half of the 12 000 inhabitants. Its core businesses are fibre optics, medical services, catalytic converters and consumer products (it is the world's largest producer of cooking equipment such as Pyrex). After losing $15.2m in 1993, the company decided that action was required to cut costs but, due to the reliance of the inhabitants of the local town on the company, the chairman refused to sanction BPR until he was certain that the local economy would not be ruined in the name of the company. Corning were in the fortunate position to be able to learn the lessons from numerous companies who have attempted BPR only to find results fall far short of expectations. Instead of imposing BPR on the workforce, it asked the employees to run the initiative for them, ensuring that they had easy access to senior management. Regular meetings were held to explain to employees what BPR entailed and several videos were produced. Employees could articulate their worries and criticisms via e-mail in anonymity. Change was evolutionary rather than revolutionary, building upon already established continuous improvement programmes. Although there was fierce discussion in churches and local bars and a survey that concluded that 33% of the workforce were suffering from stress (with only 12% coping with it) the BPR initiative has avoided massive redundancies as the company only shed 10% of the workforce. In 1994 the company produced results of a sales increase of 19% from 1993 and net profits of $281m. Even though these results are not of the order of magnitude subscribed by Hammer and Champy, etc., they demonstrate that BPR can work in a socially responsible way.

Case studies: The situation at the national level

As we shall see later in our description of the Construct IT initiative, many of the concepts of strategic IT planning are evident in what is currently underway in the UK. Other national examples are also worthy of discussion in outline.

Singapore

Singapore has long recognised the beginning of the information age. The National IT Plan (NITP) was launched in 1986 by Brigadier General (Res) Lee Hsieng Loong at the opening of Singapore Informatics, 1986. It provided a blueprint for an action programme that calls for the exploitation of IT to develop a strong export-oriented economy. The NITP has seven interactive building blocks, namely IT industry, IT manpower, information communication infrastructure, IT application, coordination and collaboration, IT culture, and climate for creativity and enterprise. Figure 4.8 illustrates these building blocks.

The construction industry falls under the IT application programme, under which three strategies were developed:

- build up an extensive information infrastructure to enhance the usefulness of IT;
- promote IT as a means to increase productivity and business competitiveness; and
- extend the public sector computerisation programme to encompass users in the private sector.

The NITP set out a challenge to all sectors of the economy to develop their use of IT resources in a nationally coordinated manner. The construction sector was asked to take up this challenge – to work towards the success of the NITP – and see how the three strategies proposed under the NITP could be adopted for the industry. Considering the fragmented nature of the construction

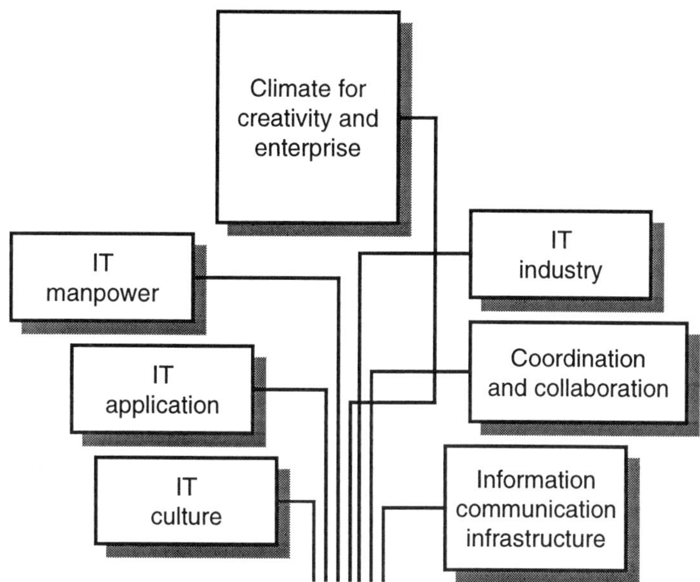

Figure 4.8 The National IT Plan (Singapore)

industry, and the peculiarities of the construction process, this was a challenging task.

The government, since the launch of the NITP, has taken steps to provide the necessary support to the seven building blocks in the NITP. An excellent information communication infrastructure already exists, and IT is almost a culture within an industrial climate that is conducive to creativity and enterprise. The National IT Plan has more recently been followed by the IT2000 initiative whereby the National Computer Board has undertaken a series of sectoral studies and overall assessments of the strategic plans necessary to make sure that as a nation Singapore is using IT effectively.

Although it is nearly ten years since this initiative was launched, electronic data interchange (EDI) soon become a reality with the start of TradeNet, a nation-wide EDI network for the trading community, in January 1989. A similar network for the health care industry, MediNet, and LawNet for legal services were developed and implemented. Other networks were planned for other industries. A civil service electronic network to serve the various government departments was also developed. While the use of IT in construction is currently lagging behind many other sectors, the potential is considerable. The Singapore government's initiative in the development of the Singapore Land Data Sharing Network (SLDSN) is a step in the right direction. This supports a full range of land management solutions rather than simply automating maps. It is important to understand the role of private sector organisations within such electronic networks and how they will operate within the larger information system within the construction industry. Such networks must communicate with other information systems used in the construction process, and not be stand-alone systems.

The sectoral study for construction and real estate, of the economy-wide IT2000 project, was part of the overall national strategic IT planning exercise which resulted in a report having a series of recommendations. These include a value-added computer network that was given the name COREnet. This stands for the Construction and Real Estate Network. This was intended as a textual and graphics data exchange mechanism that would embrace information services and EDI services. The construction and real estate sector IT2000 report also contains recommendations for a number of CORE Integrated systems and CORE Business systems to support various construction activities. It is clearly the most ambitious planning for IT use in the construction sector in Singapore to date.

Thus two conclusions emerge. First, that many have been concerned with isolated and individual developments that are causing islands of information to form. The IT2000 work is the first national-level, industry-based attempt to counter this. The second conclusion, that also applies to the IT2000 initiative to some extent, is that most previous developments have concentrated on computerising or automating the current manual processes rather than considering the opportunities that IT offers to improve the effectiveness of the industry from a strategic and process analysis point of view. However, national IT planning in Singapore is more advanced at an infrastructural level than in many countries.

A similar picture is beginning to emerge from New South Wales (NSW) in Australia. The Federal system in that country means that initiatives happen at both a national and regional level. The National or Federal Department of Industry, Science and Tourism has recently undertaken a strategic review of the construction sector. It has identified the strategic potential the sector offers to the rest of the Australian economy and to international trade. It advocates improved use of IT in the process and plans to implement this on major Federal demonstrator projects. At a State level, the NSW Department of Public Works and Services published a government Construction IT Strategy in April 1998 (New South Wales Government, 1998a) as part of a broader Construction Strategy (New South Wales Government, 1998b). These both paint a picture of a future NSW construction industry in which e-commerce, project databases, communications networks and virtual project teams are the norm. It is encouraging industry to adopt new ways of working and seeks to implement the process by sponsoring pilot demonstrator projects, setting performance goals, adopting the principles of the strategy as requirements on government contracts, undertaking industry benchmarking and encouraging training in IT.

Finland

The situation in Finland is quite different. Here a long tradition of interest and expertise in construction IT research has concentrated on gaining a lead in the technology. Product modelling has formed a focus to this and the RATAS model provides a product model definition that underpins many IT applications as an aid to standardisation and which has been widely referred to internationally. RATAS evolved into TELERATAS in translating the principle to an implemented information system. Much of the work in Finland has again been government-led through the public research laboratories. These have however been very successful in gaining and maintaining the input of key private companies who have then implemented the recommendations for standardisation. The range of ongoing government projects in Finland is also impressive. Finland has the world's highest per capita participation rate in the Internet. The recent VERA programme is a 200 man year research initiative to improve information networking in an improved construction process. Finland sponsors construction IT research to a greater extent than most other developed economies that have more than 10 times its population and GDP.

5 The scope for IT in construction

Martin Betts and Andy Clark

"You cannot do business without IT."
– Group IT director, UK contractor

Strategic IT planning techniques can be applied to many different operating levels of the construction sector. There are many examples of strategic use of IT by international construction companies including Dillingham Construction in the US for refurbishment modelling in response to new legislation following the San Francisco earthquake of the 1980s. Other companies have made strategic use of IT in the purchasing area based on exploiting their large market share. There are also national plans for IT use in construction in Singapore, Australia and the UK.

- Planning techniques can help construction companies identify strategic opportunities for IT.
- IT must be managed very differently for companies that view it as a support or strategic resource.

Some companies have found major strategic advantage through using IT in response to local market opportunity. These include Dillingham in the US who have used visualisation techniques linked to project planning to other differentiated services in refurbishment modelling.

"Ask yourself why there is always a man with a pneumatic drill on every building site? Because there has been an error in the transfer of information and you have to do it all over again. It happens every day, ... even in our company."
– Project manager, major French contractor

5.1 Introduction

In this chapter, we examine further the extent to which the strategic use of IT is being applied in construction. We review major IT applications within construction and examine them in the light of the strategic issues regarding IT introduced in Chapter 4. This chapter provides an appreciation of the techniques from Chapter 4, more fully in a construction context – we revisit some of the examples introduced earlier and review them in this light.

5.2 Information intensity in the construction sector

> "Although a trend toward information intensity in companies and products is evident, the role and importance of the technology differ in each industry. Banking and insurance, for example, have always been information intensive. Such industries were naturally among the first and most enthusiastic users of data processing. On the other hand, physical processing will continue to dominate in industries that produce, say, cement, despite increased information processing in such businesses."

This quotation from Michael Earl (1989) is of some significance to us when considering construction. Certainly you might assume that the reference to cement production is generally applicable to construction. Despite devoting significant resources to IT and the economic context being appropriate, construction, as a sector, is not making IT critical to its operations. This could be a result of the apparent physical nature of the construction industry. Yet, as we saw in Chapter 4, the information intensity of an activity must be considered in terms of both the process and the product. The grid in Figure 5.1 shows this distinction and the place of different industries within it.

An important point to consider is that within an industry, parts may occupy different segments of the grid. The construction sector operates at many levels with a wide range of sizes and types of organisation. It has an enormous range of projects in terms of size and complexity. The sector also draws on skills and resources of a highly varied nature. Much of the assembly process and materials supply parts of the construction process are low on information in product but not in process. Yet, design and management in construction are information intense in both. The

Figure 5.1 Information intensity matrix (source: Porter and Millar, 1985)

		Information content of product: *Low*	Information content of product: *High*
Information content of process	*Low*	Oil refining	Banking, newspapers, airlines
	High	Cement	

complexity of the different parties, and the communications between them, confirm this. The authors of the grid also acknowledge that over time there is likely to be dramatic movement from one quadrant to another. This movement will be by individual firms and by industries as a whole. Gaining advantage over competitors by taking different positions within the grid is an important strategy. Given that most construction enterprises may be in the low–low quadrant, this may enable other firms to find scope for competitive advantage in construction. This could be by differentiating themselves from the majority. In other parts of the construction sector, organisations may be in the high–high quadrant, making IT strategically important in other ways. On the whole, we must conclude that the information intensity of much of the construction sector's process and product is relatively low, although there is scope for increase and for parts of the sector to be considered distinct from the rest.

Earl has also commented on sector level assessment of IT impact with a classification of metaphors for the strategic context of IT in different sectors. Table 5.1 illustrates this in a model we referred to earlier in Chapter 4.

In this classification, Earl recognises the difference between sectors. He stresses that there is movement between these stages within sectors. Note that within a sector, enterprises and parts of the sector are at different stages. Although the cement production activities may remain delayed, there is evidence of some construction enterprises having reached the drive stage. We also can see that some parts of the construction sector may soon approach the dependent stage. This requires strategic thinking about future use.

A related classification has an analysis of the current and future impact of information systems (IS) within various types of companies. Figure 5.2 shows the four quadrants that an enter-

Table 5.1 Sector framework for IT (Earl, 1989)

Strategic context	Characteristic	Metaphor
IT is the means of delivering goods and services in the sector	Computer-based transaction systems underpin business operations	Delivery
Business strategies increasingly depend on IT for their implementation	Business and functional strategies require a major automation, information or communications capability and are made possible by these technologies	Dependent
IT potentially provides new strategic opportunities	Specific applications or technologies are exploited for developing business and changing ways of managing	Drive
IT has no strategic impact in the sector	Opportunities or threats from IT are not yet apparent or perceived	Delayed

Figure 5.2 Position of information systems in various types of companies (source: McFarlan, 1984)

Strategic impact at present	Strategic impact in the future: *Low*	Strategic impact in the future: *High*
Low	**Support** Large chemical company Large process industry manufacturer	**Turnaround** Insurance broker Medium-sized grocery chain
High	**Factory** Major airline	**Strategic** Major bank 1980 Large insurance company Major bank 1981

prise may occupy with examples of a company's placements based on case studies. This, again, is the same scaling framework referred to in Chapter 4. The strategic quadrant represents companies that are critically dependent on the smooth functioning of the information system activity. IS planning is closely integrated with corporate planning. Turnaround quadrant companies also need IS planning but mainly for long-term and future performance. General management guidance to IS development is critical to these companies. Factory quadrant companies do not require critical IS planning with much less senior management involvement. IT is critical to middle management in day-to-day operational issues. Finally, support quadrant companies do have some IS constraints to corporate planning but not of a significant nature. The insurance and banking cases shown within this grid illustrate that different enterprises within an industry can be at different stages in their level of strategic IT use. They show that the position of enterprises can change even over very short periods of time.

A number of companies and industry groups are and will remain appropriately in the support and factory boxes. Technical changes have been so sudden in the past several years that the role of a company's IS function needs re-examination to ensure its placement is still appropriate.

Construction enterprises planning strategically for IT use for competitive advantage are doing this now as they approach the turnaround stage. Up to now, the construction sector has been preoccupied with IT support applications.

A case for this can be made now because of the particularly dynamic nature of construction at present, as we described it in Chapters 1 and 2. The operating environment is subject to economic, technological and social effects from within and between countries. Greater deregulation, international cooperation and globalisation are beginning to have an impact on construction. The clients of the industry are increasing in their sophistication particularly with regard to quality, environmental impact and general expectations of performance. The construction industry is changing more now than it has ever done before and this in itself gives rise to strategic opportunities in the future.

The strategic potential grid can be used to show an enterprise's strategic position. This arises from the opportunity to use IT for competitive advantage in the future. Figure 5.3 shows a classification of types of enterprises by this grid. It is of greater significance to construction enterprises from an offensive and defensive viewpoint. The construction sector may have been safe from the viewpoint of strategic potential until now. It is becoming increasingly clear that, as time passes, construction will need to use IT strategy offensively.

Based on these arguments, opportunities for the widespread,

Figure 5.3 Potential for strategic use of IT resources (source: Ives and Learmonth, 1984)

		Quality of information systems resources *Low*	*High*
Value adding potential	*Low*	Safe	Explore
	High	Beware	Attack

strategic use of IT in construction appear imminent. All operating units within the construction sector must consider where their opportunities lie now to take advantage of the competitive situation.

"The absence of any pattern of expenditure on IT, irrespective of how it is measured and compared, suggests that IT is not being used by the [construction] industry in a strategic way."

The above arguments and the quotation suggest that the construction sector has been left behind by the changes in IT thinking. IT is increasingly playing an integral part within the strategic planning agenda of many large contractors and consultants. It seems that there may soon be a significant move towards strategic IT planning in the construction sector, although the developments are at an early stage. Many pilot studies are currently under way as companies realise the significance of IT and attempt to utilise it effectively. However, these developments will only become standard practice in years to come and are as yet unused by the vast majority of the industry. We should evaluate these more fully before attempting to answer the question posed by this chapter, i.e. whether the concepts of IT strategy are becoming more widespread in construction.

5.3 IT – the general picture

The overall picture is of a range of different approaches being taken in different countries and environments. These are taking place within organisations in response to external influences. Different countries have alternative approaches to the applications of IT.

5.3.1 Within organisations

It seems that IT is mostly being applied by individual participants in the interest of their short-term gains or cost savings. Little attention has yet been paid to the overall interests of the client, let alone to the wider environmental, social or cultural context and implications. IT must be exploited and applied to cover these wider issues and the interests of all the stakeholders, diverse as they may be. All construction organisations, from individual enterprises to professional institutions and trade associations and ultimately to national policy makers, must act to overcome these specific difficulties in construction if IT use is to become more strategic.

The professional bodies and government agencies may have important roles to play in different environments, from co-ordination to developing and maintaining data and information bases for their members. This would avoid duplication of work and also help small to medium enterprises who cannot see justification in spending money on large databases in isolation.

5.3.2 External influences

All types of organisations need to also see IT become critical to their current and future activities. This can be enforced by a central agency making IT necessary to certain key activities where construction organisations interface with them. In the UK, to ensure that the newly introduced coordinated project information system was adopted by the industry, all organisations working on projects commissioned by the former Property Services Agency (which was a major and critical source of public projects) were required to follow the procedures and documentation recommended by the Coordinating Committee for Project Information (1987). This could easily be extended to embrace IT applications of Coordinated Project Information. In Singapore, the Construction Industry Development Board envisages making the submission of tenders for public sector projects in magnetic form mandatory as one of the means of making IT critical. This is also part of the thinking behind the NSW government strategy in Australia where specific IT capabilities for consultants and contractors are set as pre-qualification requirements for government construction work in a three-stage IT improvement plan. Such approaches by a central agency may have a role in different economies.

5.3.3 Alternative approaches to IT in various countries

To harness the strategic advantage that IT use would provide, an effort appropriate to the local circumstances will be necessary. One such approach would be a government-led, industry-wide effort. At this level, information is potentially shared and communicated extensively, across professional boundaries and throughout the construction process. The strategic effect of such an effort is likely to be improvements in the external competitiveness of the industry as a whole. This is the most beneficial and broadest level at which IT strategy can be applied. Any IT development at this level would require the understanding, participation and support of all professions, enterprises and government agencies concerned. Such a national, industry-wide approach is what is now being undertaken in the Construct IT initiative in the UK, the VERA programme in Finland, the CORENET project in Singapore, and the Making IT Happen strategy in Australia.

In the UK, professional bodies are taking strategic steps. At this level, information is being shared and systems are being developed for all subscribing members of a professional body for their individual benefit. Strategically, this is leading to improvements in the productivity of the professions but is not addressing the issue of their external competitiveness. A further complication may be that the professional bodies lack a co-ordinated approach. In each body, developing solutions for disparate operating frameworks and systems, the industry again faces the hindrance of a lack of integration and more fundamentally a lack of standards.

The Japanese approach is for strategy to be the responsibility of construction enterprises. The leading organisations have developed their activities to embrace all stages of the design and construction process, thereby internalising the integration problem. Strategically this is leading to improvements in the productivity and internal competitiveness of the enterprises and, because of their size and strength, this effectively means the external competitiveness of the industry is improved as well.

In Singapore (as in many other newly industrialising and developing countries), the weakness of the professional institutions and the small size of corporate enterprises combined with the strength of the central agencies make a nationwide initiative more appropriate. In the UK, the professional demarcations were much stronger and the bodies themselves more active. In Japan, the corporate culture makes strategic initiatives by enterprises the only feasible alternative. Each of these alternatives has implications for the form and impact that IT strategy takes and makes.

5.4 The five-level framework

In deciding how strategic our use of IT has been up to now, we should consider any examples of strategic management and process analysis having been applied to IT implementation at one of the five levels of the construction sector already described. The same five-level framework can now be used with some of our earlier examples but this time in terms of the role of IT.

5.4.1 National level

In Japan, the Information Network System is a national plan for an IT network that has important implications for construction. Research in Scandinavian countries, aimed at coordinating IT research and infrastructure for the construction sector, is also of consequence. Some more ambitious and extensive plans for national exploitation of IT for strategic advantage come from Singapore. Within a national climate of long-term IT planning, specific national programmes exist for public and private sectors. Within this context, the plans for an integrated national communications network for the strategic advantage of the home construction industry appear exciting. These plans explore the issues of linking organisations throughout the industry and of building fundamental inter-organisational information systems. In the US, the issue of computer integrated construction, and work to bring it about, has received the attention of the National Bureau of Standards. As the competitive strategy of nations becomes more important, this level of the applications framework will assume greater importance. Up to now, few nations have viewed strategic opportunities for IT use by their construction sectors. There is considerable scope for greater IT policy-making in the construction sector in many parts of the world. The New South Wales government in Australia has recently followed this route with its state IT strategy for construction, Making IT Happen.

5.4.2 Professional level

A dominant approach in the UK was formerly for strategic management and process analysis by individual professional bodies. At this level, shared information and IT systems development is for the individual benefit of subscribing members of professional bodies. Examples that we have already described come from both the UK architecture and quantity surveying (QS) professions, the latter in two ways. In both, the profession is justifying its role as more than performing a quantity take-off. These initiatives are of strategic importance as defensive strategies in guaranteeing profession survival in a climate of technological progress and deregulation. IT forms a key part of the RICS Challenge of Change (1998).

5.4.3 Enterprise level

Most historical construction sector examples of strategic IT use are within enterprises. This is where the greatest commercial pressures exist. For our first example we can look at one of the world's largest engineering construction firms. The M.W. Kellogg Co. based in Houston, Texas, executed US$6.2 billion of contracts in 1985 with 92% of their work outside the US. They ranked amongst the top five of the top 250 international contractors. They had stated aims in response to the growing competitive pressures of the global construction industry. These were to compete through their computer systems. They sought a high state of integration for efficiency gains as part of a dual service differentiation and cost leadership strategy.

A second US constructor, Tishman Construction Co. of New York, went further in using IT for strategic advantage. Manhattan's largest general construction managers built extensive site and office networks linked by modems and networks. They have further offices in Los Angeles, Florida and Chicago. They installed modems and links to their systems to allow clients access to office and site information. They invited clients to access on-line information regarding project status, pending items, cost and schedule reports. The strategy is one of product differentiation with speed of delivery as a key measure. They add the cost of the client link to the project bill. There are considerable benefits in Tishman's owner links and the importance the organisation gives to IT research should be stressed.

The more recent case example of Dillingham Construction described more fully later in this chapter, using modelling based CAD to exploit niche markets in seismic foundation and retrofitting engineering, is a further example of strategic IT use. The

Skanska case is a further example of strategy using IT to exploit buyer power.

Other, longer established historical cases include Grupo ICA, a leading Mexican constructor. This involves the use of site microcomputers for strategic opportunities. Japanese contractors have also been prominent in this area. Taisei's plans to become a knowledge-intensive firm and Shimizu's for an integrated CAD/CAM/CAE system as part of its Engineer Constructor strategy are noteworthy, as are Taisei's plans for integrated CAD systems and corporate networks. Whilst discussing the enterprise level, remember that we do not only refer to contracting companies but also to design and consultancy practices, trade subcontractors, building material and component suppliers and the many other specialist organisations that make up the industry. By doing this we can find the most dramatic and often quoted example of strategic IT use by a construction sector enterprise that we described earlier. This is the OTISLINE example described in Section 4.3.3.

Further examples of material and component supplier implementation are also to be found. Competitive buyer power forces could justify a decision support system of German building materials suppliers. An example of a simple stock control device has potential to be much more. The United Building Co. of Alaska has benefited - during an economic downturn - from implementing a point-of-sale stock control system. These are both examples of the result of lower cost, value chain analysis.

There are details of the strategic exploitation of IT research and development by Owens-Corning Fiberglass Corporation. The US home insulation enterprise allows builders and designers the free use of energy efficiency evaluation software in exchange for guaranteed incorporation of insulation products. This is a new service made possible by IT tied to the continued use of an existing, highly competitive product and exploits supplier power.

In the UK, many major construction enterprises are looking to strategically exploit IT. These include Alfred McAlpine with their use of modelling software for stadium design and management, Bovis, with their supply chain management systems, and Laing with their Integrated Engineering Solutions initiative.

Smaller US construction firms also exploit IT in different ways. KNC Corporation in Tennessee aims for cost efficiency with IT systems and plant and equipment analysis programs in particular. Combustion Engineering Incorporated of Connecticut is another firm where IT is intrinsically linked with a major corporate strategy rethink. We therefore have many examples of strategic IT use by different types of large construction enterprises from all over the world using a number of strategic management and process analysis techniques. This should convince us of the opportunities for other enterprises who may exploit opportunities similar to those shown above or others appropriate to them.

5.4.4 Project level

This is a level of application of strategic management and process analysis absent from many manufacturing, financial services, airline and health service industries that others have studied. Yet, the project is the basic construction industry operating level. There is scope for innovation and competitive advantage at this level. This is therefore a level of strategic opportunity that may only exist in project-based sectors. Much of the current technology-based innovation in IT in construction is pitched at a project level.

This is a fertile ground for new applications of IT to emerge. IT can create new businesses and revolutionise industry structures. Further exploration of this application area is necessary besides exploring opportunities related to current professions and enterprises. This is if the full potential of construction IT is to be uncovered. The process redesign initiatives of BAA described in Chapter 3 are examples of a more fundamental exploration of the strategic possibilities of process change that are substantially enabled by IT.

5.4.5 Product level

Transformations have taken place in products due to requirements for information about the product. For example,

> "...convenient, accessible information and service procedures are important buyer criteria in consumer appliances."

In an industry where even as-built drawings are rare, the scope for improvements in the information content of buildings is large. There is an opportunity here for any number of the current professionals and enterprises, or new ones. All have the chance to use IT to provide information about constructed buildings, after completion, for use by clients and facility and maintenance managers during the occupancy phase.

In addition, given the information intensity of both processes and products, we must consider the built facility. In doing so, we build a link with a further important area of current concern in construction – intelligent buildings. The construction industry has traditionally concentrated on information within the process of construction only. The emergence of intelligent buildings is a primary force for change in strategic management and process analysis. The success of intelligent buildings in Versailles, France, New York's World Trade Center and Tokyo's NEC Supertower shows the appeal of intelligent buildings which is leading to aggressive marketing. This may be an example of a new product made possible by IT as part of a focus strategy.

A particular feature of the construction sector is that its primary product consists of many different sub-products (in the form of materials and components) which are then combined and assembled into a building. The implication of this is that in thinking about IT strategy at the product level, we must also include building materials and sub-assemblies, many of which have a growing information intensity from which strategic IT opportunities are likely to follow.

5.5 National plans for IT use in the construction sector

5.5.1 IT in construction (Singapore)

We have already described the national initiatives taking place in Finland, Singapore and Australia in the case studies at the end of Chapter 4. We now return to the more construction specific activities in Singapore to get a fuller picture of strategic IT activities at the level of the construction sector. Participants in the construction process generate and use information that is specific to the project at hand, obtained from the enterprises they are associated with, or available in the public domain to the industry at large. As a project develops, the information base grows very quickly. Each enterprise participating in the project depends on a large information base that it has developed over a period of time through its involvement in other similar projects. Public and private agencies and enterprises that support the industry, such as materials manufacturers and suppliers, the planning authority and professional institutions, provide information to the participants. Thus, within construction there are conceptually and in practice three distinct information systems:

- project information system;
- enterprise information system; and
- industry information system.

The project information system (PIS) is used during the planning, design and production phases, and should be retained throughout the life of the building. Information is drawn from an enterprise information system (EIS) and an industry information system (IIS) to develop the PIS. Thus the EIS and IIS could be considered to be general information systems supporting the development of the project-specific PIS. Since several enterprises participate in a construction project, the PIS has to communicate with several EIS. Since at present, the IIS does not necessarily have to be a unified source of information, the PIS may have to communicate with several sources of information which, in our conceptual model, represent the IIS.

Figure 4.3, discussed as an example of an impact model in Chapter 4, shows the way in which these three information systems can conceptually combine within a single integrated national IT framework. It also shows the channels of communication among the PIS, EIS and IIS. All enterprises participating in a particular project have direct on-line access to the PIS. The PIS is shown as a separate system outside all other EIS because it is not the property of any one enterprise. Traditionally the architect's enterprise maintains the PIS, but within an electronic data exchange environment it might, more typically, simultaneously reside with all participating enterprises in the form of a distributed project database.

Any changes made to the project design or additional information produced will immediately be registered in the files of the PIS held by other enterprises. The PIS is conceptually shown as an independent system with which all enterprises interact. Enterprises do not normally communicate with each other except in the context of a specific project. Communication between two enterprises is therefore shown via the PIS. Enterprises may also communicate with each other directly during a project but this would lead to the PIS being amended. The PIS therefore results in greater interaction among two or more enterprises.

This framework was presented to leading members of the

Singapore construction industry representing public and private sectors and all relevant professional bodies. A broad consensus supporting such a national plan was obtained. Singapore thus has a blueprint for strategically exploiting IT in construction as a nation. The work in strategic management and process analysis for IT does not finish with such a framework, but it could be argued that it cannot start without it. Implementation of such a strategic framework should progress with the advancement of existing network systems and the launch of new databases. All such efforts should be made with reference to the national framework, the details of which should be further developed and defined through continuing industry-supported research. The situations in Australia and Finland are also more coordinated and advanced than many, as we saw in Chapter 4.

5.5.2 The UK Construct IT Initiative

The Construct IT Centre of Excellence is an industry-led initiative committed to the advancement of the industrial capabilities of the UK construction industry through a comprehensive programme of innovation. It began in the summer of 1995 and is a network of more than 70 leading companies, institutions and universities with an administrative base at Salford University. Its aim is to implement and continue the development of a construction industry process improvement strategy for the UK, focusing on information technologies as an enabling mechanism for industry change.

The centre involves an active partnership between:

- industry, including all construction parties, IT and communications companies;
- experts of leading universities in the UK and the international scientific community; and
- research sponsors, including the Department of Environment, Transport and the Regions, the Engineering and Physical Sciences Research Council and the Innovative Manufacturing Initiative.

The centre was set up by industry to address the recommendations set forth by a series of major recent studies and reports such as the Latham Review (1994), the *Technology Foresight: Report of the Construction Sector Panel* (Office of Science and Technology, 1995) and the *Bridging the Gap: An Information Technology*

Strategy for the UK Construction Industry report (DoE, 1995). The plan of work of the centre is based on the understanding that improvement in the performance of the construction industry is contingent upon the improvement of processes enabled by the dynamic collaboration of industry and the research base and effective support from information technologies.

The programme of work of the centre makes a major contribution to technology policy and innovation management in construction and research into such interdisciplinary fields as computer integrated construction. It also addresses the motivational and skill requirements involved in these technological changes.

The centre delivers practical results which include:

- a long-term national research and technology development programme;
- short-term pilot innovation projects including demonstrators;
- benchmark analyses;
- a programme of postgraduate and continuing education;
- IT management tools;
- networking between major UK players;
- communications and international links; and
- a coordinated industry response to the millennium date change problem.

The centre recognises the need to design different communication strategies to meet the different needs of industry and the research community. The industry members are encouraged to discuss the results of the programme of work of the centre through seminars, workshops and members' meetings. Dissemination to a wider industry audience is achieved through business publications. Moreover, the high quality of the work is ensured by targeting publication in reputable academic and industry journals and by securing continuous state-of-the-art international information and input.

The incorporation of an evaluation procedure also ensures the high standard of the activities of the centre. Evaluation includes examination and assessment of the mode of action and effectiveness of the policies of the centre. The evaluation procedure results in closer cooperation between the commissioning institutions and the evaluated organisations, founded on improved mutual awareness.

The development of the Construct IT Centre of Excellence is the outcome of a timely coincidence between the emergence of an industry-led initiative aimed at coordinating research in construction information technologies and the publication of a government strategy for information technology in construction for which an implementation body was required.

The industry-led collaborative venture includes clients, consultants, major contractors, leading universities, and a number of other leading companies representing the IT and communications industries. A key part of the plan of work of this group included the definition and implementation of a national IT strategy in construction.

5.6 Current state of IT-based business process applications in construction

With larger resource bases, process strategies are currently the domain of large contractors. John Laing, Kyle Stewart, Balfour Beatty and Tarmac are currently involved in process-based initiatives either with external consultants or internal project teams. Other organisations with a strong interest in process redesign include John Brown Engineering and the Bucknall Group. Their reasons for adopting these methods differ from the case study of BAA (p. 76) due to their market position. It is interesting to note that the organisations are drawn from a range of sectors in the industry, suggesting that process thinking has a wide applicability. The initiatives are driven by the need for:

- a strategic advantage;
- internal efficiency;
- better client service; and
- international competitiveness.

Their processes are designed to affect the operations of a company as opposed to the operations of the project network, so that process structure reflects business needs, which may include project needs.

At the time of writing it is difficult to assess accurately the success of BPR in the construction industry. A reasonable timescale to expect substantive results for BPR is over two years; it would be unfair to expect serious improvements at this stage. Moreover, due to the conservative nature of the construction industry, the required change could take longer and should be expected to take longer.

5.7 Strategic goals and hindrances

5.7.1 Progress towards strategic goals

From the analysis in this chapter, it cannot be argued that the construction sector has fully entered the IT era. However, it is possible to see that by a number of criteria, this situation may change in the future. Some individual developments that have taken place in different parts of the world do show some isolated progress towards the IT era. It is also possible to identify what change is necessary to bring the IT era closer on a broader front. This may happen through commercial pressures or through the industry as a whole recognising the need for change. In some economies, it may be necessary or desirable for this process to be helped or speeded up by some form of central direction or control.

There are many limitations with the current use of IT in construction. The IT tools that are used are standalone systems, many of which were originally intended for the engineering design/production process. Few of the tools pay due regard to the nature of the construction process and the alternative form of design approach and construction technology. Most IT products have been developed as commercial ventures for an industry that is fragmented into tightly defined roles and with short time horizons. For these and other reasons, there are only isolated examples of the strategic use of IT at present. The implementation of the new IT, particularly that based on CAD, knowledge-based systems and robotics, will be a difficult task since the management and dissemination of this technology in a fragmented industry requires careful and strategic management and process analysis at the industry and professional levels. The issue is not just that of implementing IT as a productivity tool in individual organisations and on specific tasks. For strategic competitive advantage, IT has to provide an 'environment' in which the computer functions as an intelligent tool for communication, coordination, book-keeping, problem-solving, automation, and decision-making.

For IT to become a strategic weapon for construction organisations, the various enterprises and agencies should do more than just consider its contribution to internal productivity. Internal and external competitiveness will be important strategies and the role of research and development by the enterprises themselves, or through their trade associations or professional institutions, is important. Central agencies may offer financial or other incentives to such research or initiate pump-priming projects of their own.

It is essential for all levels and functions of management to embrace IT. The professional institutions may influence this issue by identifying IT awareness and management skills as an important component of continuing professional development programmes. Educational institutions must ensure that relevant courses are provided for both professional updating and for initial undergraduate education.

The great diversity of stakeholders in construction, many of whom do not see beyond short-term gains, is such that advancing the strategic use of IT is a difficult management problem. Alternative catalysts and primary movers are possible, depending upon the local environment. The effective coordination of the very wide range of organisations and activities which will be necessary will require strategic frameworks and detailed tactical planning by a central agency. Such an agency will also need to consider the problems of technological matters.

The need for automation in construction is caused by increasingly sophisticated quality and time demands from construction clients and by the shortages in skilled construction workers in some parts of the world. One aspect of the drive towards automation has been an attempt to apply IT. Integration has been used as a goal in our efforts to use IT for automation. The aim of this chapter is to demonstrate how the strategic IT planning frameworks used in other business sectors apply to construction. Research and practice requires tactical and strategic developments. It is argued here that much of our preoccupation to date has been with the tactical issues that relate to technology.

We believe that the strategic goals for IT in construction are integration, communication and automation. These provide the vision necessary for focused development and implementation efforts. They also offer the opportunity for the continuous improvement necessary to the success of computer integrated construction (CIC). With regard to CIC they provide some answers to the question of how integration might be achieved. Other questions that may be considered include who will achieve it, what will be integrated, when and why?

CIC is a goal towards which many are working, but we must recognise the variety of problems that inhibit its achievement. These include matters of technology and how to integrate. This is the area where we are concentrating our efforts at present. We appear to be developing sophisticated and extensive answers to this question. We must also recognise that where, why and who are important questions to ask. This chapter attempts to provide some possible answers to these other, currently unanswered, questions.

5.7.2 Hindrances to strategic IT use in construction

The sectors where IT appears to have made most dramatic impact are where other major forces for change exist. Some examples of these are quoted (Earl, 1989): deregulation of airline and financial services; the need for global survival in automobile manufacture and textiles; and the structural changes that are affecting retailing are all cited as forces which are leading some organisations in some sectors in some countries to use IT as a strategic response.

A further prerequisite for strategic use, and a symptom of an organisation having entered the IT era, is that the technology must have permeated all functions and levels within an organisation. Early uses of data processing (DP) were restricted to distinct DP departments and for specific number-crunching applications. Organisations that have more fully embraced IT are now in a situation of having a range of hardware and software technologies being used by diverse groups of people and for a variety of different tasks and activities.

In construction this is far from the case. In many construction organisations IT is still being used by IT specialists, for discrete applications and it would seem only by staff at the technical levels. This situation is borne out by studies that have observed that IT is still too restricted to administrative functions of an accounting nature or for highly specific and technical functions within the construction disciplines.

Due to the complex network of relations that contribute to a construction project, standards remain a key issue. The lack of a critical mass to impose standards has resulted in parties being unwilling to make technological advances, especially with respect to communication technologies. Fears of security also militate against communication technologies – there is a fear that e-mail and EDI could provide access to systems containing sensitive information. Firewalling systems offer protection but cases of hackers entering Pentagon systems do nothing to dispel fears. It must be pointed out that it takes a hacker of rare genius to gain access to a firewalled system and many cases of hacking are exercises in publicising the skills of the hacker, not genuine attempts to gain sensitive information.

On this basis it must be concluded that the construction sector is a long way from having reached the IT era, or being strategic in IT use. A significant change is needed in the variety of levels of management that are using the technology and the diversity of functions to which it is being applied. For this to be possible, all construction organisations will need to reconsider their size, organisation structure, recruitment policy and their education and training activities in the light of their requirements in the IT era. Not all appear to have done so as yet. The solution to the problem is only partially to be addressed by developments in standards and technology.

There are many isolated and individual examples of enterprises and projects where IT is being used for improvements in internal productivity. The area where construction organisations appear to be failing to exploit IT strategically is for their internal or external competitiveness. Few can offer a new or improved service by virtue of their IT use and even fewer consider new ways of managing and organising themselves or developing new businesses. From this it can be concluded that on the basis of IT being a potential strategic weapon, the construction sector falls short of having fully arrived in the IT era.

International case studies

Dillingham Construction

A major international case study being included in this chapter is of Dillingham Construction. Dillingham is a US building contractor based in Pleasonton, California. Its main building company within the group is 49% owned by Shimizu of Japan. It has some 320 personnel and more than 2500 site staff.

Within the holding company is an operating business unit called Dillingham Builders Inc. which has around 20 staff. Within this business unit there is a technical services manager who has been responsible for implementing an IT-based innovation.

The group has its own MIS department in central services along with office services, human resources and an executive department. The MIS department of five or six people work with the technical services manager but the distinction between central services and innovation is clearly understood. The President of Dillingham Builders is now personally spearheading this innovation and keeping it separate from the MIS department.

Dillingham became involved with this innovation because of their involvement with the Center for Integrated Facility Engineering (CIFE) at Stanford University. They were interested in pure technological experimentation, got involved, liked what they saw and sponsored a specific research project for an application that they felt had strategic business possibilities. They ended up employing the researcher, keeping him separate from the main IT department as a problem solver. He was being protected from the top.

This case involves a specialist application of 4D modelling software. They were using 3D CAD and applying time as a fourth dimension to visualise the sequence of construction to their clients. This has been described by the company as a research-based IT experiment that was successful. It is described as an experiment in that no-one had fully shown the benefits of 3D CAD modelling for the particular type of application that Dillingham were interested in. They took between three and four months to build a 4D CAD model for a large public hospital project that was started in March 1994.

The senior management champion who initiated the project was the head of construction for Northern California. He was widely perceived within the company as a visionary. He was personally inspired by academic visions and thought there would be benefits in applying new research to specific practical issues. In particular, he foresaw constructability benefits from seeing conflicts and clash avoidance and working towards a concurrent engineering paradigm based on a federated architecture technique. A 4D model of the San Mateo hospital was built up from existing 2D models as an innovation in parallel with existing applications and traditional benefits from IT.

The success of the innovation arose from the system going above and beyond the initial brief improving the client's reputation locally because of the benefits they were getting from the publicity that the system attracted.

They used software called WALKTHRU which was originally designed by Bechtel and is now owned by Jacobus Technologies which is in the Bentley family of companies. The original research was scaled down from real-life applications but this application was large scale. Some examples of its benefits include the fact that the concrete subcontractor designing forms was given enough visualisation to find buildability advantages.

The benefits to the hospital in showing stakeholders the implications of a retrofit project have been of much greater impact. The modelling application is very specific. For new-build projects the benefits are mainly in constructability. For existing buildings it can offer much more. The advantage of the project was also that they saw how to re-organise the building schedule to avoid ongoing operational difficulties as the new facilities were constructed.

The original application of the technology on the San Mateo hospital has led to new invitations to bid that would not otherwise have arisen and has given Dillingham the confidence to do new research. The original award of the San Mateo contract to Dillingham was not on the basis of the lowest bid. Dillingham now strongly feel that they have a competitive advantage with this technology by being able to differentiate their services. They feel they can offer services that other contractors cannot and that they have created a market for a new kind of service in doing so. There is a perception within Dillingham that both their competitors and clients realise the company has a competitive advantage and communicating this to these groups has been an important part of the innovation.

An important part of their current activity has been to interface the technology with hospital staff during the construction period. The Hospital Phase 1 remodel needed extensive planning effort that, now made, can be used to show clients what will happen during construction. This is an important part of the client's and other stakeholders' need.

Communication of the continuing benefits from this innovation are held to be key. The company has had many articles about this innovation written for US professional magazines. They have also paid for the production of a video of their modelling. Promoting their innovation has been key and has been partly achieved through the winning of a National Competition for Professional Services Marketing.

Dillingham see themselves as following a niche strategy of seismic upgrade modelling to hospitals and biotechnology facilities. They are doing this with a technology push that is meeting with a market pull being created by changes in regulations. Since the San Francisco earthquake of the early 1980s, a new state law has been passed known as the 2008 Plan. This has created the Office of State-Wide Health and Planning Department (OSHPD) which requires extensive new seismic upgrade work to existing occupied buildings in many health and biotechnology areas. This created a new market to which a technological innovation has provided a solution. Dillingham now see seismic retrofitting as a major commercial opportunity.

Since the publicity that was gained from the application of 4D CAD, Dillingham have been contacted by many new clients for possible seismic upgrade projects. One of these new clients is talking of hiring Dillingham to do 4D modelling as a separate service contract. This is part of the Dillingham business strategy to do more of this sort of work. Dillingham foresee that their company might even have a 4D modelling division in the future. Their value-added and service components have increased through IT.

Haka

Haka were a Finnish contractor increasingly influenced by the changing economic and political climate in Scandinavia and throughout Europe. This, combined with the emerging technological developments in which they retained a close interest, led them to examine IT carefully as a potential strategic weapon. In particular they examined diversification within the project life cycle as a strategic possibility. Haka's implementations and computer developments were based on using product models developed and defined by the government research laboratories and amending them for their own situation. Implementations based on these product models were used largely for a diversification strategy.

Haka saw the potential of IT to broaden their limited contractor-only service. They envisaged products and services being offered through a broader range of the project life cycle arising out of their core competence in IT based on a product modelling technology. This was very much to support the general Haka corporate strategy of becoming a networked company reacting to declining domestic markets by forming strategic alliances with developers and joint venture contractors in North America and Western Europe to exploit links in Eastern Europe. IT was seen as a key technology to support this networking approach. Haka's strategy was not able to prevent the company being taken over by Skanska of Sweden in the early 1990s following cash shortages. This is an example of the difficulty of operating a long-term technology-based strategy within the constraints of short-term survival. Skanska themselves are a further example of an interesting international case study and more details of their strategic use of IT follow in later chapters.

Shimizu

Shimizu are also increasingly looking to IT as a cornerstone of their business strategies, particularly with regard to productivity improvements, quicker construction and advanced technology exploitation. Shimizu's situation is quite different from Haka – their role and product/service range are already very broad. Their strategy with their SMART (Site Management Automation and Robotics Techniques) system has been more to integrate various technologies including IT. Their aim for integration is part of a broader production philosophy whose order of magnitude productivity objectives have more to do with lean production (Womack *et al.*, 1990) than some of the other strategic models we describe here. The principles of the SMART system are described by Yamazaki (1995).

The evidence of these different approaches should not confuse or worry us but merely show that a range of options are available depending upon where an organisation is starting from, where it is trying to get to, and its local regulatory and economic context.

6 Strategic information systems planning techniques

Marjan Sarshar, Martin Ridgway and Martin Betts

"IT must be driven from the top."
– Managing director, UK contractor

To effectively realise new business strategies as an organisation, and to effectively exploit modern IT, information systems (IS) strategies are required. There are a number of techniques for IS strategy planning. These include soft systems methodology, information engineering and process innovation. These methods all take quite different approaches. Which to use within your business will depend on the criteria you have for an IS strategy; the skills and resources available; the nature of the deliverable you seek and personal taste. An international multi-disciplinary design practice chose to use information engineering (IE) in preference to strategic systems management and process innovation (PI) for reasons that are detailed in the chapter.

- Converting a business strategy and plans for strategic use of IT into an IT strategy requires an IS strategy.
- Criteria for selecting a strategic information systems planning (SISP) methodology include: ease of use, fit with company policy and resource availability.
- Alternative SISP methodologies include SSM (soft systems methodology), IE (Information Engineering) and PI (Process Innovation).
- Each methodology has strengths and weaknesses that make them appropriate to different situations.

"When you give people too much information they get fed up, stop using it, and end up throwing it away."
– Accounting director, French contractor

6.1 Introduction

In the previous chapters, we have developed an understanding of why strategic management is so critical to modern business thinking and of how the concepts apply to construction. We have then explored the strategic role of IT in modern business thinking. Much of what has gone before is concerned with changing our thinking and improving our understanding. Whilst critical to accepting and beginning a process of change, in itself, this does not help us do things differently. This chapter on strategic information systems planning (SISP) techniques takes us further. Having identified strategic initiatives that a business can take which IT can support, the next step is to develop an IS strategy and then implement this as an IT strategy. This chapter and Chapter 7 concern themselves with these two practical issues.

Work addressing an industrial problem formed the background to the writing of this chapter. A large multi-disciplinary firm needed to develop an IS strategy. The previous information strategies of the firm were mainly technology based. The strategy documents generally discussed which hardware was needed and how to standardise on this hardware. A few application packages were also mentioned in the strategy. But it was unclear why these choices were strategic to the company, and how the company was to reap competitive advantage as a result of implementing them.

The company had a new IS manager who wanted a review of their systems and technologies. He was keen to understand which were strategic to the company and what the company priorities were. He was also seeking new solutions which would fit into his company's business culture and processes.

The first step was to identify a systematic methodology which was tried and tested and could be used in the company to develop an SISP. One senior manager had been on an MBA course and suggested the use of the soft systems methodology (SSM). The IS manager wanted a second opinion to ensure this was the right choice. They discussed their decisions with leading academics and asked for an investigation of what was possible. It quickly became clear that numerous approaches and methodologies exist, but little practical evaluation of any of them was available.

The university academics examined general characteristics of SISP methodologies and selected criteria for their objective comparison. They then chose three named methodologies and examined their strengths and weaknesses by interviewing two experts in each field. The findings provide some initial insight for methodology selection.

As we can see elsewhere in this book, throughout the 1980s large contractors and consultants were willing to increase overall expenditure on IT, as money was available to explore its potential. However, this period was followed by a recession in the UK in the early 1990s when budgets were cut and expenditure on IT was restricted. This often gave rise to isolated spending on IT/IS, only where it was considered to be absolutely critical to short-term survival, and without regard for a long-term coherent strategy. Many departments in construction organisations are now faced with dealing with legacy systems, the presence of which detracts from other IT/IS investments that the firm may wish to take.

It is also clear that there is still a fear within the industry that investment in IT/IS is risky due to the danger of backing the wrong technology or standards, and a fear concerning the difficulty of keeping investments up to date. Failure to deliver some of the benefits promised by IT/IS projects in the past, coupled with an ongoing difficulty in quantifying the return on investment, have also contributed to the general reluctance within the construction sector to invest in IT/IS in recent years (Department of Environment, 1995).

Against this background of legacy applications, lack of IT/IS investment, and the fear of getting it wrong, there is a need for a more focused approach for developing and implementing a strategy to deal with the information resource. One technique that can help to overcome these barriers is to adopt an existing 'tried and tested' methodology for the strategic planning of information systems. Not only should a proven methodology help get over the initial hurdle of commencing a strategic plan, but it should also help to ensure that the resulting strategy is both effective and sustainable.

In the mid-1980s a number of methodologies appeared for analysing IS from a strategic perspective. Because of the variety it may be costly to find an appropriate methodology, and even more costly to use an inappropriate one. There are few guidelines for appraising and selecting methodologies. This chapter is based on a case study of a large multi-disciplinary design firm who was attempting to select an appropriate methodology.

The philosophies behind the methodologies vary, and there is no common method of classification. One technique is to classify the various methodologies according to their approach, e.g. data-driven, process-driven and technology-driven, etc. In selecting the methodologies it was decided that contrasting methodologies should be examined. Three methodologies were selected, one which was process driven, one which was data driven and one which was systems driven. These provide a spectrum of SISP methodologies.

The three SISP methodologies selected by the firm for consideration were:

- information engineering (IE), which is predominantly data driven;
- soft systems methodology (SSM), which is driven by the analysis of human activity systems; and
- process innovation (PI), which is driven by processes and technology.

In the following sections we introduce SISP methodologies and some criteria for their comparison. Then we briefly describe these three selected methodologies. This is followed by the results of semi-structured interviews with two experts in each methodology, to assess perceptions of the strengths and weaknesses of each.

6.2 Strategic information systems planning

A SISP plan does not just deal with IS in isolation, but sets out a strategy for dealing with the whole of the information resource, from data and raw information through to individual systems and technologies, together with their relationship with the complete business environment.

There are many definitions of what SISP is, most of which refer to individual methodologies. However, we consider the following definition by Wilson (1990) most helpful:

> "An IS strategy brings together the business aims of the company, an understanding of the information needed to support those aims, and the implementation of computer systems to provide that information. It is a plan for the development of systems towards some future vision of the role of IS in the organisation."

A lack of a coherent ISP produces a number of problems for construction companies.

- Business opportunities are missed. Companies may even be competitively disadvantaged by the IS/IT developments of others. Systems and technology investments do not support the business objectives and may even become a constraint to business development (Betts, 1992).
- Integration of information is a key factor in construction IT (Department of the Environment, 1995). Currently, technical solutions for integration are emerging. However, construction companies are still unable to reap the benefits of integration (DoE, 1995). As we have seen in earlier chapters, implementation of integration is a strategic issue, which requires effective management of data, systems, resources and people (Ward and Griffiths, 1996). ISP provides guidelines for accomplishing implementable information integration.
- Without an ISP, technology strategy is incoherent and unimplementable within the business. This leads to the selection of incompatible options. Large sums of money are wasted attempting to fit things together retrospectively (Ward and Griffiths, 1996).
- Lack of shared understanding and agreed direction between users, construction managers and the IS specialists leads to conflict, inappropriate solutions and a misuse of resources. This in turn leads to construction companies not finding 'value for money' in IT applications (Construct IT, 1997b).

Earl (1993) established that many SISP methodologies focus on some or all of the following four target areas:

- aligning investment in IS with business goals;
- exploiting IT for competitive advantage;
- directing efficient and effective management of IS resources; and
- developing technology policies and architectures.

These target areas need to be incorporated into an overall strategy plan, which carefully maps out the utilisation of the information resource.

In order to help understand how to plan IS, it is necessary to see how an IS strategy fits into an organisation, and its relationship with other elements of the organisation. Figure 6.1 depicts the relationship of IS with IT and business strategy (together with brief notes of the components and the function of each element), which helps to view IS strategy in the overall business context.

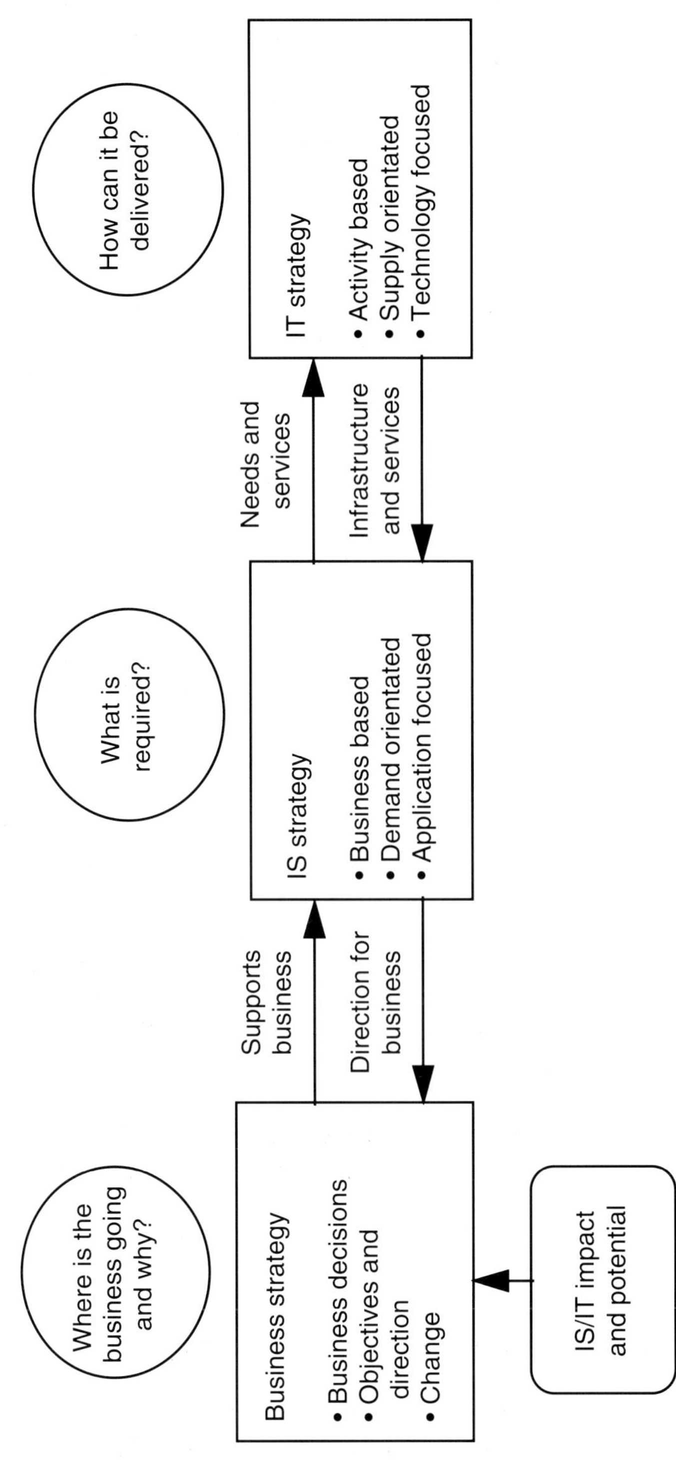

Figure 6.1 The IS/IT/business strategies and relationships (source: Ward and Griffiths, 1996)

We have considered business strategy, and how it may be reconsidered in the light of the strategic role of IT, in earlier chapters. Here we are concerned with how IS strategy can be planned. Chapter 7 is concerned much more with practical issues of implementing an IT strategy.

An attempt to plan IS usually raises many different business issues, due to the close relationship with business strategy. These issues are frequently complex, and, coupled with the difficulties of a rapidly changing business and technical environment, make the task of providing a SISP exceptionally daunting. It is not surprising, therefore, that many organisations elect to use an existing methodology to help provide a strategy for planning their IS, rather than developing an in-house planning technique. Tozer (1996) also lists some more general reasons why a methodology is used for SISP.

When considering adopting a methodology it is, therefore, essential that an organisation should consider what a methodology offers. A methodology may:

- range from a fully fledged product detailing every stage and task to be undertaken, to being a vague outline of basic principles contained in a short brochure;
- cover widely differing areas of the SISP process, from high-level strategic problem-solving to detailed tasks and procedures;
- cover conceptual issues, or detailed physical design procedures, or the whole range of intermediate stages;
- be designed for a specific environment, or be an all-encompassing general-purpose methodology;
- be potentially usable by anybody, or only by trained specialists;
- require an army of people to perform all the specified tasks, or it may not even have any specified tasks; and
- may include CASE tools.

Some methodologies are purchased as a product, others are available by purchasing a licence, others are obtained through a contract for consultancy work, some come as part of the purchase of the CASE tool, and some by a combination of the above.

6.3 Criteria for methodology selection

Researchers have proposed various guidelines for comparing methodologies. Avison and Fitzgerald (1995) propose three main categories of rationale behind the selection of a methodology for the development of IS, which are equally applicable as criteria for adopting a SISP methodology:

- a better end product;
- a better development process; and
- a standardised process.

A better end product refers to the output of the SISP process, and criteria for assessing this will depend on the objectives for carrying out a SISP project. For example, Earl (1993) lists the top five objectives from a survey of 21 companies as:

- aligning IS with business needs;
- seeking competitive advantage from IT;
- gaining top management commitment;
- forecasting IS resource requirement; and
- establishing a technology path and policies.

Criteria for assessing the end product of a SISP project will depend on how the importance of the above criteria are ranked, and how successfully the methodology is able to recognise and prioritise these needs.

A better development process, in terms of SISP development, refers to the benefits of tightly controlling the process and output (deliverables) at each stage. This results in improved management and project control. It is usually argued that productivity is enhanced with the use of a methodology, i.e. needing fewer resources to obtain the same result. A standardised process relates to the benefits of having a common approach throughout an organisation. This means that a base of common knowledge and experience may be built up and shared amongst the members of the organisation.

Chatzoglon and Macaulay (1996) suggest six criteria to be considered when considering the selection of a SISP methodology:

- ease of use;
- support manuals available;
- design: based on sound principles and valid assumptions;
- support and integration of management activities such as cost estimation, scheduling, team organisation;
- quality of techniques appropriate for the specific environment employed; and

- quality outputs accompanied by verification and validation capabilities.

Other criteria which may also be used for selecting a SISP methodology include:

- users already familiar with a particular methodology;
- no other alternative available;
- company development policy; and
- particular criteria fit best with a certain methodology.

It was found in practice that there are other reasons for selecting particular methodologies including head office policy, historical reasons, literature study and a preferred supplier, all of which suggest that selection is not only influenced by the characteristics and advantages of the methodology, but simply reasons of history or familiarity.

Considerable practical problems are encountered when attempting to compare SISP methodologies.

- Methodologies are not stable; they are moving targets, continually evolving and developing. Therefore, a version problem exists and it is difficult to know which version of the methodology is being applied in a particular situation, or which is the latest version.
- For commercial reasons, the documentation is not always published or readily available to people or organisations not purchasing the methodology.
- The practice of the methodology is sometimes significantly different to that prescribed by the documentation.
- Consultants using the methodology often interpret aspects of it in quite different ways.
- Problems of undertaking comparisons due to terminology, in particular, the use of different terms for the same phenomena.

A common approach to methodology comparison attempts to identify idealised features followed by a check to see whether different methodologies possess the features or not. The implication being that those that do possess them or at least score highly on a features rating are 'good', and those that do not are 'less good'. The set of features must be chosen by somebody, and thus are subjective.

In writing this chapter two experts in each of the selected

methodologies participated in a semi-structured interview to identify the strengths and weaknesses of each method. The results are presented in the next sections along with an overview description of each methodology.

6.4 Introduction to soft systems methodology (SSM)

The soft systems approach results from work carried out at Lancaster University and first formalised in Checkland's publication, *Systems Thinking, Systems Practice* (1981).

The methodology was developed through action research where general systems ideas were applied in organisations. As such, it is not specifically designed for SISP. It has, however, been used as a 'front end' for hard IS development and was first suggested as a method for formulating a coherent information strategy by Checkland and Scholes (1990). It is further noted that using SSM for information analysis of purposeful activity models often leads to useful discussions of both information requirements and, on a broader scale, information strategy in an organisation.

6.4.1 SSM overview

SSM, as first formalised by Checkland (1981), comprises a seven-stage model which is summarised in Figure 6.2. The model has developed over the years but for the purpose of SISP, it is best to consider the conventional seven-stage model which underpins the methodology, and is still commonly in use in all the developed forms.

In summary the seven stages are as follows:

(1) a problem situation is acknowledged in its unstructured form;
(2) the problem situation is expressed (this often entails the development of a 'rich picture');
(3) root definitions of relevant systems are provided;
(4) conceptual models of the purposeful activities are developed;
(5) the conceptual models are compared with the real world;
(6) changes are proposed – these should be systematically desirable but culturally feasible; and
(7) action to improve the situation is taken.

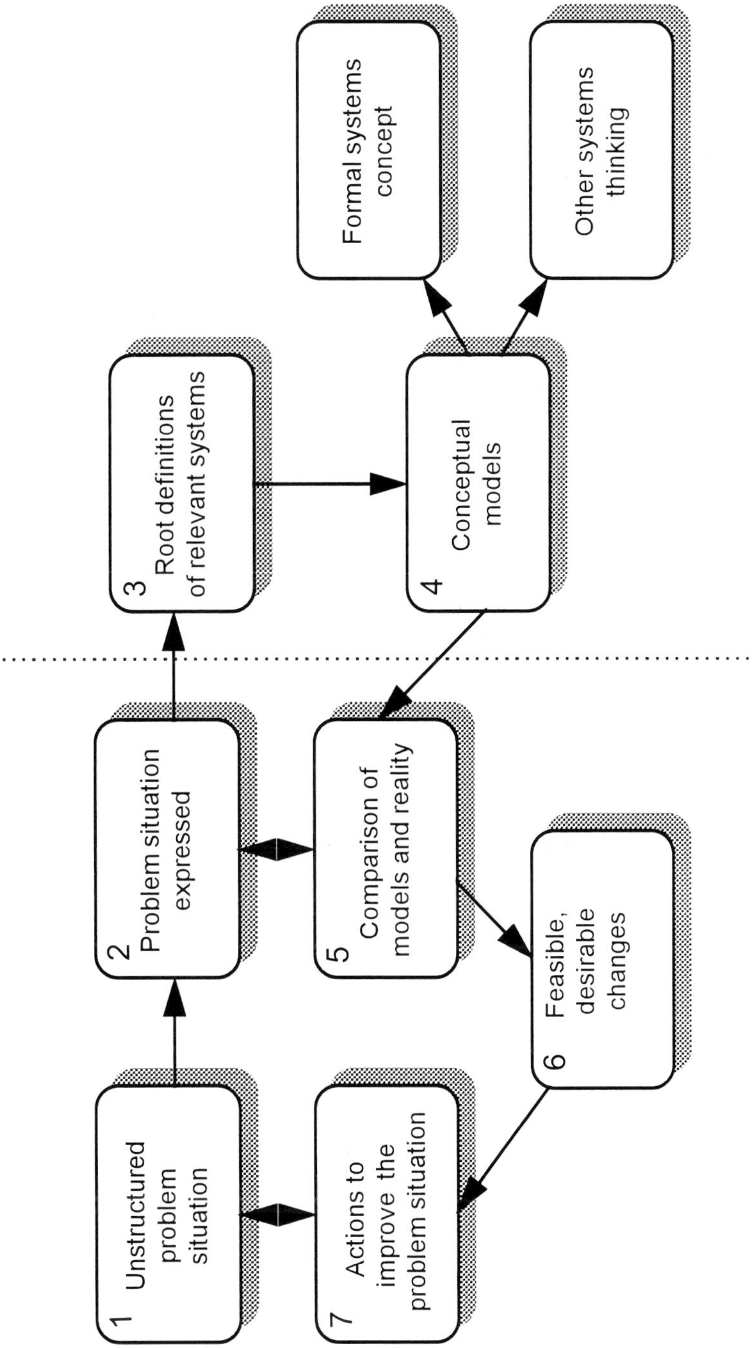

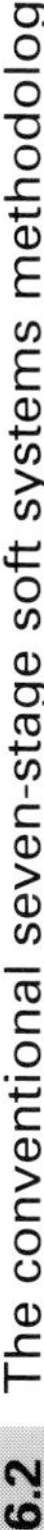

Figure 6.2 The conventional seven-stage soft systems methodology

To the left of the dotted line in Figure 6.2 are the 'real world' activities involving people in a problem situation, whereas stages to the right of the dotted line involve activities concerning thinking about the problem situation in a systems way.

Stages 1 and 2 are about finding out about the problem. The unstructured view gives some basic information from the individuals involved. The application of criteria (Checkland, 1981) gives some structure to the expressions of the problem situation and in Stage 3 the analyst selects from those views which he or she considers give insight to the problem. Stage 4 is to do with model building, that is what the systems analyst might do (as opposed to what the system is – the root definition). There must be one conceptual model for each root definition. Stage 5 compares the conceptual models from Stage 4 with the root definitions formed at Stage 2. This comparison process leads to a set of recommendations regarding change, and Stage 6 analyses these recommendations in terms of what is feasible and desirable. Stage 7, the final phase, suggests actions to improve the problem situation, following the recommendations of Stage 5.

Soft systems methodology was not originally conceived as a SISP methodology. As the methodology developed, SISP was seen by SSM practitioners as an area where the methodology could make a contribution. Checkland *et al.* (1996) illustrate this is by looking at a typical case study where SSM was used to provide a SISP.

6.4.2 Strengths and weaknesses of SSM

Two senior academics participated in semi-structured interviews to examine the strengths and weaknesses of SSM. They each have more than fifteen years of academic as well as consultancy experience in the area of SSM.

Strengths

- Simple to use.
- Essentially a top-down methodology which helps to achieve buy-in from top management, and therefore easier to control the management expectations.
- Size of planning team is relatively small due to the non-complex nature of the methodology and the majority of the analysis being prepared by users.
- Graphical output is easy to understand, i.e. 'a picture paints a thousand words'.

- Time taken to establish and define what the real problems are, for example, an organisation could have already evolved an IS strategy, but the real problem could be that it is not properly formalised or it may be that there could be difficulties with implementation of a strategy.
- Does not make assumptions about an organisations structure, unlike IE (information engineering) methodology which assumes a hierarchical structure.
- Closely identifies human issues, and the human resources issues such as training needs can be readily identified.
- Conceptual models generate alternative views of the problem situation, which can be tested against the real world problems, then adopted, enhanced or discarded, as appropriate, whereas other methodologies may not provide such rich alternatives.
- All the processes are modelled, and in particular their interactions, to provide the 'big picture' for IS needs.
- Many of the skills used by SSM for SISP are transferable to solving other enterprise problems.
- Involvement of users encourages teamwork and ownership of the problem.

Weaknesses

- SSM was not designed as a SISP methodology.
- Despite the methodology being 'simple' some users have difficulty in embracing the problem-solving philosophy of SSM, and would prefer a more rigid set of rules to follow.
- There is no CASE (computer-aided software engineering) tool support.
- Difficult for the novice to manage SSM, and external support is usually required from an experienced practitioner. This is particularly important when first describing the SSM philosophy in attempting to achieve management buy-in.
- There is a need to involve a considerable number of users with the SSM. Some organisations find it difficult to commit to such a process.
- The output fails to address implementation, although this may be perceived as a project management issue and not a methodology issue.
- Does not provide an overall data architecture for the organisation; SSM focuses on processes rather than data. This could lead to a lack of systems integration in practice.

- Does not provide a technical architecture.
- The final output is often criticised for being too simple, and management may often comment that ' I could have told you that before!' Therefore expectations require careful management.

It is noted that SSM provides a suitable basis for SISP despite the fact that it was not designed as a SISP methodology. The main strength of SSM is its approach to exploring the problem situation in detail, but its weakness is that it focuses on processes at the expense of data, and does not provide a complete systems architecture for the organisation.

6.5 Introduction to process innovation (PI)

Process innovation is a methodology which ties business process redesign (BPR) with IS/IT. Importantly, PI is an approach to information systems which takes into account strategic aspects of BPR (Avison and Fitzgerald, 1995).

Towards the end of the 1980s, the idea of redesigning or re-engineering business processes gained popularity. Two articles on the topic of process innovation and the role of IS/IT (Davenport and Short, 1990; Hammer, 1990), are credited for identifying the basic principles in this area, and in 1993 Davenport published his methodology, *Process Innovation: Reengineering Work Through Information Technology*. We have commented on these process redesign developments in earlier chapters.

The methodology is presented to organisations as an opportunity to dispose of outdated rules and assumptions underlying business practices, and to redesign processes, including those performed by computer systems, which are responsible for underperformance. This redesign is enabled by technology. Therefore, PI is not sold as a SISP methodology *per se*. However, the outcome of PI is to redesign processes and to provide a strategy for IS and IT to support the redefined business environment.

6.5.1 The PI philosophy

The essence of BPR is a radical change in the way in which organisations perform business activities, thus it is defined by Hammer and Champy (1993) as:

> "the fundamental rethinking and radical redesign of business process to achieve dramatic improvements in critical, contemporary measures of performance, such as cost, quality, service and speed."

Redesign determines what an organisation should do, how it should do it, and what its concerns should be, as opposed to what they currently are. Emphasis is placed on the business processes, and therefore the IS that reflect them and enable the change, and also encompasses managerial behaviour, work patterns and organisational structure.

Many businesses claim that IT has not yielded the productivity increases that its pundits promised, and the adoption of process methodologies aims to address the fundamental flaws in the application of IT. The main objectives for IS have been automation and process rationalisation, which have only produced minor incremental productivity improvements. We shall return to the issue of how the benefits of IT innovation are measured in some detail in Part C of this book.

As we have seen in earlier chapters, processes cut across traditional functional boundaries, forcing the organisation to analyse how products and services are treated from inception to completion. Hammer and Champy (1993) argue that managers lose sight of the overall process by working in functional areas, while Davenport and Short (1990) point out that few companies analyse performance based on process criteria. Many business processes have evolved from *ad hoc* decisions made by functional units, which were implemented to improve their own effectiveness, rather than the effectiveness of the whole process. Process redesign reverses this in practice, maximising the performance of the entire organisation by focusing on the interdependencies within and across the sometimes disparate functions that comprise the organisation.

Process structure has been subject to much debate amongst the redesign theorists. Although any company could count between 2 and 200 processes, there will be an optimum number. Davenport (1993) suggests identifying between ten and 20 core processes, because if the processes are too broad, it will be impossible to fully understand their complexity, and therefore difficult to substantially improve on them. If the processes are too narrow, BPR will only produce minor gains to overall business performance.

According to Davenport and Short (1990), IS and BPR have a recursive relationship. On the one hand, IT usage should be determined on the basis of how well it supports redesigned business processes. On the other hand, BPR should be considered within the realms offered by IS (and the supporting IT). The combination of IS and BPR presents organisations with the opportunity to change the way in which business is conducted. The increasing complexity of the environment has presented organisations with new threats and challenges. The need to maximise the performance of interrelated activities rather than the individual business functions, combined with the opportunities offered by IS, has meant that a new approach to the coordination of processes across an organisation is necessary to achieve competitive advantage. Consequently organisations need to ensure the close alignment of IS with business strategy, through strategic information systems, the latter being derived in the context of the changing competitive environment. This is the line of thinking that has underpinned many of the earlier chapters in this part of the book.

6.5.2 The five stages of PI

The following steps of process redesign are crucial to the success of re-engineering processes with IT and were mentioned in Section 4.4. It is worth repeating them here in the context of this discussion. They are to:

- develop business vision and process objectives;
- identify the processes to be re-engineered;
- understand and measure existing processes;
- identify IT levers which will help push the changes; and
- design and build a prototype of the new process.

Develop business vision and process objectives

In this phase, the organisational strengths and weaknesses need to be identified, along with an analysis of the market and the opportunities it provides. A knowledge of the innovative activities of competitors will also be useful. It must be noted that a business vision will only come as a result of creative thinking of business executives and others.

Identify the processes to be re-engineered

At this stage the major processes are identified, along with their boundaries. The critical processes of the organisation are considered for IT-enabled redesign. Processes which are of high impact, of great strategic relevance or presently conflict with the business vision in some way are selected for consideration and a priority attached to them. It is unlikely that they can all be redesigned in parallel. There may be somewhere between ten and 20 processes identified for innovation.

Understand and measure existing processes

Processes cannot be redesigned before they are understood. The present processes must be documented. This will help communications within the group studying the process. It will also help to understand the magnitude of the change and the associated tasks. Understanding the existing problems should help ensure that they are not repeated. It also provides measures, which can be used as a base for future improvements. For example, measuring the time and cost consumed by process areas that are to be redesigned can suggest initial areas for redesign in a process. However, although designers should be informed by past process problems and errors, they should work as if in virgin territory, otherwise processes will be tampered with, rather than redesigned.

Identify IT levers which will help push the changes

The accepted view for designing IS is that the business requirements should be determined before considering IT solutions. This methodology proposes that an awareness of IT capabilities can influence process redesign and should be considered at the early stages.

Design and build a prototype of the new process

In this final stage, the process is designed and the prototype built through successive iterations. Design comes from a review of the information collected in the first four stages. It is suggested that the design team consists of key process stakeholders as well as personnel with IT expertise who can debate possible design alternatives.

6.5.3 Strengths and weaknesses of PI

We interviewed one senior academic and one senior practitioner to examine the strengths and weaknesses of PI. The following is a summary of the strengths and weaknesses of PI, as identified by these interviews.

Strengths

- The methodology mainly relies on internal resources to examine the organisation's own processes. External support is essential, but only in an advisory capacity.
- Encourages the use of CASE tools available to support analysis and implementation.
- Suitable for use with rapidly changing industries and technologies.
- Addresses external competitive and technical environments.
- Analyses strengths and weaknesses of current processes.
- Aligns business strategy with process vision.
- Identifies priorities for projects.
- Provides an opportunity for radical business change and competitive advantage.
- Focuses on customer needs.

Weaknesses

- Not conceived as a SISP methodology.
- Success is dependent on drive for PI coming from top management.
- Planning and analysis take a long time.
- Documentation does not thoroughly describe steps needed to implement the methodology.
- Does not provide an overall data architecture and technical architecture for the organisation.
- Difficult to secure top management commitment due to radical nature of the methodology.
- High failure rate: more than 50% of business process redesign projects are thought to be unsuccessful in terms of their ambitious initial objectives.
- Not sympathetic to human issues.

The main strength of PI is the radical examination of the business and its processes to enable an overall systems architecture to be designed to support the redesigned business processes. The emphasis on the external environment is a further strength, which helps to focus on customer needs and achieving competitive advantage. The main weakness is the high risk involved in undertaking this approach to SISP as, historically, many PI initiatives fail to deliver full benefits.

6.6 Introduction to information engineering (IE)

Information engineering (IE) originated in the USA and Australia (Finkelstein, 1989). It is a software development approach, which covers all the aspects of the life cycle, and fundamentally, a methodology, which considers that data are far more stable than the processes which use them. Thus it is proposed as a methodology which successfully identifies the underlying nature and structure of an organisation's data as a stable basis from which to build IS. Every system has at its heart information, or more correctly data, and the IE approach is to arrange the data in a structured framework and store them in a data bank which provides an easy means to access the data. That is not to say that IE does not take into account processes – clearly it recognises that processes have to be considered in detail in the development of IS – and balances the modelling of data and processes as appropriate.

IE has been explored in a construction context several times in the past (Sarshar *et al.*, 1993, 1994). There are several methodologies for information engineering. One of the most commonly used proprietary versions is that of Sterling Software (formerly Texas Instruments) which utilises the Composer (formerly Information Engineering Facility) CASE tool, and this version of the methodology is considered here.

The methodology commences with a top-down approach, and begins with a top management overview of the enterprise as a whole. In this way, separate systems are potentially related and coordinated and not just treated as individual projects. This enables an overall strategic approach to be adopted. As the steps in the methodology are carried out, more and more detail is derived and decisions concerning which areas to concentrate upon are made. Based on the overall plan, the business areas to be analysed first are selected and then a subset may be chosen for detailed design and construction. This approach to the management of complexity in IS requirements of an organisation is termed 'divide and conquer', and it is illustrated by Avison and Fitzgerald (1995).

The objectives and the focus change as the methodology progresses, with each stage having different objectives, although the overall objectives remain consistent. Progress is controlled by measuring whether the objectives have been achieved at each stage, not by how much detail is generated.

IE is divided into four stages:

- information strategy planning (ISP);
- business area analysis;
- systems planning and design; and
- construction and cutover.

The objective of an ISP is to construct an information architecture which supports the overall objectives and needs of the organisation. This is conducted at the enterprise level.

6.6.1 ISP tasks

Developing an ISP is a joint activity of senior general management, user management and IS staff. It involves the performance of four tasks: current situation analysis, executive requirements' analysis, architecture definition and documenting the ISP as detailed in Figure 6.3. These four tasks are now explained below.

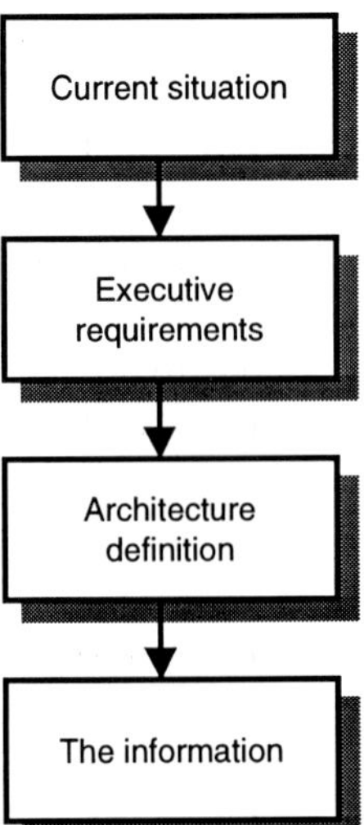

Figure 6.3 The four ISP tasks

Current situation analysis

This is an overview of the organisation and its current position, including a view of the strengths and weaknesses of its current systems. This overview will include an analysis of the business strategy, an analysis of the IS organisation, an analysis of the technical environment, and a definition of the preliminary information architecture (data subject areas, such as customer or services, and major business functions).

Executive requirements' analysis

Here, managers are provided with an opportunity to state their objectives, needs and perceptions. These factors will include information needs, priorities, responsibilities and problems. This also involves the identification of goals of the business and how technology can be used to help achieve these goals, and the way in which technology might effect them. Critical success factors (CSFs) for the overall organisation are identified and these are also broken down into CSFs for individual parts of the organisation.

Architecture definition

This is an overview of the area in terms of information (the identification of global entity types and the decomposition of functions within the subject areas described in the preliminary information architecture in the current situation analysis above), an analysis of distribution (the geographic requirements for the functions and the data), a definition of the business systems architecture (a statement of the ideal systems required in the organisation), a definition of the technical architecture (a statement of the technology direction required to support the systems including hardware, software, and communications facilities) and a definition of the information systems organisation (a proposal for the organisation of the IS function to support the strategy).

Information strategy plan

This includes the determination of business areas (the division of the architectures into logical business groupings, each of which could form an analysis project in its own right), the preparation of business evaluations (strategies for achieving the architectures, including migration plans for moving the current situation to the desired objective) and the preparation of the ISP itself (a chosen strategy including priorities for development and work programmes for high priority projects).

6.6.2 Strengths and weaknesses of IE

Two senior practitioners participated in semi-structured interviews to identify the strengths and weaknesses of IE.

Strengths

- Comprehensively documented, describes all the steps to be followed for implementing the methodology.
- Supported by powerful CASE tools.
- May be undertaken by employees of the organisation without the need for external consultants.
- Takes into account organisational goals and strategies.
- Appraises current IS and provides migration proposals.
- Identifies current external, technical and competitive environments.
- Involves top management for helping align IS with business strategy.
- Information gathered during ISP can be used at later stages, including developing individual IS and generating programming code.
- User involvement may be kept to a minimum or it can be very participative.
- Determines priorities for projects by analysing business objectives/critical success factors.
- Provides hardware, software and communications architecture.
- Focus on data helps identify data architecture, data stores/databases.
- Final output document is a very helpful communication tool.

Weaknesses

- The planning exercise takes a long time.
- It is a complex methodology, which requires time for top management to understand and gain commitment.
- It is not sympathetic to 'people issues'.
- Does not sufficiently involve the end users.
- Output is very rigid.
- Relies heavily on a hierarchical organisation structure, while in the 1990s many organisations are moving to flat, process-based structures.

- Processes are subservient to data.
- Does not encourage teamwork.
- Based on relational technology, and cannot easily be extended to object-oriented technology.

The main strength of IE is the completeness of the methodology. It takes into account data, processes and their integration, and provides an overall architecture for the organisation. One of the main weaknesses is the rigid nature of the output, which assumes that bespoke systems will be developed, without considering the use of component or 'off the shelf' systems.

6.7 Comparison of the methodologies

Table 6.1 summarises the above findings. The characteristics listed in this table are drawn from an earlier section of this chapter. The list of criteria has then been expanded, based on the interviews with the experts. The assessments are also based on the interviews with the experts. Some issues which have not been addressed in Table 6.1 include the following.

- The duration of the SISP project. Business managers and users lose interest in a long IS project. Today's businesses are changing rapidly. The results of a SISP project which may take one or two years are usually too old for the business.
- Project risks have not been adequately addressed. This is due to lack of sufficient empirical evidence. Project risks can arise from the inherent weaknesses of the methodology, or the way they have been implemented in a particular company. However, a good SISP methodology should have sufficient project management mechanisms to predict risks and take adequate precautionary measures.

Case study: IS needs of a multi-disciplinary design practice

This section of the chapter examines the organisational characteristics of a large multi-disciplinary practice. Many of these characteristics are common amongst UK and international construction firms. The case study can provide some insight into the required features of an SISP methodology for construction generally. The findings in this section are based on the following information provided by the company:

Table 6.1 Comparison of SISP methodologies

Characteristic	IE	SSM	PI
Aligning IS with business needs	Excellent	Moderate	Excellent
Seeking competitive advantage from IT	Moderate	Poor	Excellent
Gaining top management commitment	Moderate	Good	Very poor
Forecast IS resource requirements	Excellent	Poor	Moderate
Establish technology paths and policies	Excellent	Poor	Moderate
A standardised development process	Good	Poor	Poor
Simplicity of use	Moderate	Excellent	Poor
Simplicity of development techniques (ease of use without consultant support)	Moderate	Poor	Poor
Support manuals available	Excellent	Poor	Very poor
Design: based on sound principles and valid assumptions (leading to low-risk implementation)	Moderate	Moderate	Poor
Support of project management activities	Excellent	Poor	Poor
Quality outputs accompanied by verification and validation capabilities (including CASE tool support)	Excellent	Very poor	Good
Efficient use of human resources during SISP project	Good	Moderate	Poor
User involvement	Moderate	Good	Moderate
Focus on systems integration	Excellent	Very poor	Poor
Focus on customer needs	Moderate	Poor	Excellent
Opportunity for radical business change	Good	Poor	Excellent
Focus on human resources issues	Moderate	Excellent	Very poor
Creating business process focus	Moderate	Moderate	Excellent

- a set of business documents;
- the IT health check questionnaire (Chapter 13), completed by 65% of the company's senior managers; and
- a series of in-depth interviews with nine senior managers. These interviews were specifically set up as part of this research.

The top-level business document is the company's mission, which states their aim:

> "To be Europe's leading building design practice, delivering excellence in design and service through partnership."

Furthermore, the aims of the business are also articulated in a company brochure, as follows:

> "• To create fine work, appreciated by clients, users, public and peers.
> • To advance the art of creating the built environment, giving better service and added value, and perfecting multi-discipline methods to do so.
> • To sustain and develop a network of offices serving regional and international clients.
> • To earn sufficient margin to invest in the continuous improvement of our people and process.
> • To be a continuing practice, transcending individuals."

From this point onwards one would expect to see developed business plans, which would typically include high level features, such as quantified objectives, strategic plans, goals and critical success factors. These would serve as a single reference point for communicating the business directions and priorities, both internally and externally. However, the company does not have any such document. This has an impact on formulating strategic plans for functional and infrastructure activities of the company, such as IS provision.

Other findings from documentation, questionnaires and interviews are summarised in Table 6.2.

The selected methodology

It is essential that the adopted SISP methodology is sympathetic to the organisational issues, needs and characteristics of the company

Table 6.2 Company characteristics of the case study

Documentation	No single business plan. Disjointed documentation forms the basis of the business strategy. Single IT strategy document provides a framework for IT support. Business mission and aims are articulated. Critical success factors not documented.
IT health check questionnaire	IT is critical for core competencies and meeting client needs. Confusion as to whether IT is a support tool, or whether exploited strategically by the business. Conflict between value adding properties of IT and financial issues dominating IT investment. Involvement of users, IT managers, senior managers and facilitators is critical to success.
Senior manager interviews	Business planning style is *tactical*, i.e. focus on the next 12 months. No targets or performance measures for *strategic* planning period, i.e. 3–5 years. Planning and forecasting are project driven, not market driven. Partnership style of management results in slow decision-making process. Organisational culture is extremely cautious towards change. Critical success factors are diverse and unclear. Company aim to achieve competitive advantage by providing value-added services. Confusion regarding the responsibility and processes for information systems planning. Immediate priority for a management information system to assist project costing, forecasting and marketing.

in order to ensure success. Therefore, the following criteria, established through the analysis of the company's culture and management, should be satisfied by the selected SISP methodology:

- strong techniques for establishing business strategies;
- secure commitment from senior management;
- facility for graduated change to information systems to reflect slow decision process/cautious attitude to change;
- involvement of users, IT managers, senior managers and internal/external facilitators;
- management of expectations and good quality output;
- establishment of future vision and overall information architecture;
- identify information needs and priorities;
- focus on customer needs; and
- clarify IS/IT performance and investment criteria.

These requirements can be compared against the objectives of an SISP and the characteristics of the selected methodologies. The lack of a clearly defined business strategy is perhaps the single most important factor which will influence the selection of the SISP methodology. In this regard, the IE methodology offers formal structured techniques for establishing measurable business objectives, critical success factors, performance targets, information needs and priorities, and an overall information architecture to satisfy the organisation. SSM and PI do not contain such structured techniques for assessing the business strategies, and would not be able to deal with this issue as efficiently and articulately as IE. Attention is drawn to the survey by Earl (1993), in which the highest ranking objective of SISP was found to be aligning IS with business needs; therefore the importance of adopting a methodology which is able to deal with this situation cannot be overstressed.

The second most important objective of SISP, reported in the survey by Earl (1993), is seeking to gain competitive advantage from IT. Information engineering and SSM do not have specific techniques for assessing the potential for gaining competitive advantage through IS strategy; however, seeking competitive advantage through the redesign of processes and IS is central to the philosophy of PI. However, it must be said that, due to the poor status of business planning, it is unclear what the company's strategies for achieving competitive advantage are, and therefore it is difficult to see the merits of adopting a methodology which focuses on such an issue.

A third objective of an SISP is the provision of a future vision and the techniques for providing an information architecture. This is clearly a strong point in favour of adopting the IE methodology. Information engineering analyses the data uses and data needs, and has techniques for analysing functions and data affinity, which provides the basis for an information architecture, whereas PI and SSM tend to ignore data and focus on individual processes and systems, and not on overall architecture. Therefore IE is best placed to identify potential databases, and how to enable integration through IS within the organisation.

All three methodologies are able to identify the information needs and priorities for information within an organisation. However, IE has a powerful CASE tool which ensures that information systems correspond with the business needs and strategies of the organisation. This ensures that the information systems with the greatest value to the organisation are prioritised for development.

Perhaps one area where both SSM and IE are weak is the facility to focus on customer needs, unlike PI which is very strong in this area. Not to say that this is a sufficiently strong justification to recommend the adoption of PI, but it is a point to be aware of if adopting either IE or SSM.

With respect to establishing performance and criteria for investment in IT, none of the methodologies address this issue particularly well, although they all seek to address the need for IS and IT, and highlight the associated benefits.

The need for user involvement in the SISP process is clearly recognised and advocated by the company. Soft systems methodology and IE both encourage the involvement of users (unlike PI which focuses on senior management), although this will require careful management regardless of which method is adopted, perhaps more so with SSM than IE, due to its less formal structure.

A major issue to be considered in the selection of a SISP methodology is the extremely cautious attitude of the company towards change, and related to this, the partnership style of management which results in a slow decision-making process. The cautious approach towards change will almost certainly rule out selecting the PI methodology for providing a strategic IS plan. This methodology involves a radical approach to IS strategy planning, and usually results in completely re-engineering the core business processes, and subsequently the information systems. Any attempt to adopt this methodology is likely to meet resistance from management, and if adopted it is likely to result in a 'watered down approach', thus losing the thrust of this methodology. Furthermore, the benefits of this methodology are unlikely to be realised due to the already slow decision-making processes, which would be further exaggerated due to the number and magnitude of decisions required for re-engineering projects, and it would therefore lose its impact, or, perhaps, never actually be implemented.

If PI is too radical for the company, either IE or SSM will need to be adopted. Based on the above analysis of the strengths and weaknesses of IE and SSM, and the organisational needs and issues facing the company, the more structured and comprehensive IE appears to provide a more suitable approach for establishing a strategic IS plan. However, as Earl (1993) notes, in practice, many organisations use more than one methodology, or develop 'in-house' methodologies, following the experience of using several different methodologies, by combining various features from the methodologies which have been used. However, great care must be taken not to lose sight of the objectives of the outcome of SISP for the organisation, and not to become preoccupied with the development of another SISP methodology in its own right.

In consideration of these criteria, one may conclude that IE provides the most suitable basis for establishing the company's strategic IS plan. There is little doubt however that, with careful management, some of techniques employed in SSM and PI could be used in conjunction with IE to help provide a strategic IS plan, provided always that the overall objectives are achieved.

6.8 Summary

There are many criticisms of the construction industry for its incoherent and *ad hoc* IT resourcing and spending (Construct IT, 1997b). Against this background there is a requirement for strategic planning of information resources. Companies would benefit from an SISP methodology, which would guide them through their strategy planning formulation and implementation. However, there are many different methodologies available, each with different characteristics and features. There is little evidence of a single methodology which has 'stood the test of time' to be used as an industrial standard. Choosing between the available methodologies can be a costly and difficult exercise.

The findings of this chapter support the view that there is a need for more focused research into defining an industry accepted methodology for SISP. Construction companies seeking to develop an information strategy face a difficult task of identifying the right methodology for them. There is evidence that the right methodology is influenced not only by the features of the SISP methodology, but also by the business culture of the company adopting it (Ridgway, 1998).

7 Implementing an IT strategy in practice

Derek Blundell

"People support what they help to create."
– Managing director, UK services contractor

One of the most difficult problems with improving our use of IT in construction is overcoming the implementation problem. The practicalities of implementing the sort of strategies that are described in this book are extremely difficult. Technology development happens more quickly than our ability to implement and apply technology.

- The issues involved with implementing an IT strategy embrace planning, selecting software and solutions, rolling out systems and supporting them once they are in place.
- Planning issues embrace the implementation strategy, process change, setting up the project team, seeking integration, sequencing the introduction of new systems, defining timescales and resources and managing applications.
- Selecting solutions issues embrace specifying systems, choosing between bespoke and packaged solutions, selecting suppliers and packages and software development processes.
- Issues in rolling out systems include auditing skills, preparing data, training and user consultation.
- Support issues include help desks and user groups.

"The first thing to do before choosing a system is to ask the users what they really need."
– Engineering manager, Belgian contractor

7.1 Introduction

The previous chapter has shown a range of methodologies available for developing a SISP. Assuming one of these has been followed, and a technical architecture has resulted, the next problem then becomes implementing the strategy by implementing the technology. This chapter draws from practical experience of implementing an IT strategy within more than ten business units of a major international contracting company.

This chapter has been written to give a general overview on the subject of IT implementation. It is not meant to be a detailed manual. This approach has been taken because every company has different requirements depending on its size, activities and scope of IT strategy implementation required.

Throughout this book the principle adopted is that all IT development should be business-driven. Consequently IT should play a part in and be integral with most construction industry activities and business processes.

Having outlined an overall approach to developing a business-based IT operation, this chapter on IT implementation assumes that you have:

- determined that an IT strategy is required by your company;
- developed an IT strategy based on your company's business strategy and the requirements resulting from that strategy, possibly following one or other of the SISP methodologies presented in Chapter 6;
- reviewed the company's business processes and identified the relevant business area(s) to be covered;
- determined both your information flows and your systems needs; and
- carried out cost/benefit analyses.

In this chapter we address the implementation issues of:

- planning;
- software selection;
- implementation; and
- support.

7.2 Developing the basic implementation plan

7.2.1 The implementation strategy

One of the most, if not the most important aspect of an implementation plan is that it must be business-led. It may seem obvious but this point is all too often overlooked, especially in companies with numerous divisions and sections or different operating units.

For example, the corporate centre may decide to implement an overall IT strategy but this may not always be supported by all of the constituent parts of the organisation. As a result, implementing a new IT strategy can be difficult, or almost impossible, when some of the users are against it or do not feel part of it. Consequently, an implementation strategy is vital. Factors that need to be considered are:

- business processes;
- project teams;
- systems integration; and
- timescale and resources.

As already mentioned, in developing the specification for the new system it is important for you to have reviewed the current business processes and highlighted the areas of improvement that you wish to obtain from the implementation of a new system. You also need to have mapped out the new business processes you wish to achieve.

7.2.2 Process change

New systems often result from a change in the basic process of the business, though sometimes the process can only be changed by the introduction of a new system. Whatever the case, the change should be process-led and not IT-led.

Where possible it is best to change the business process either manually or within the existing system. If this is not possible then it is important to tell the users why the process is being changed. You also need to explain that it is not the computer that is changing the process or working practice, but the business. It is also important to get the approach right before you implement the system. The change management process must not be underestimated.

7.2.3 Setting up the project team

Every project needs a sponsor who should be senior enough in the company to get things done - but not that senior that the rest of the project team feel they are unable to disagree with them or have an equal say in the various decisions the team will be required to make. The team needs to represent the various elements of the company - though it is important to ensure that it doesn't contain so many individuals that it becomes inflexible and ineffective.

Team members need to have the ability to convey progress to, and seek the advice and approval from, their peers. However, whilst it is not always possible to represent or obtain the views of every interested party within the company, the chosen team members should fairly reflect the views of all interested parties.

7.2.4 Integration

When the IT strategy consists of implementing one or more solutions, and you plan to pass data from one system to another, then you need to consider systems integration and how to achieve it. There is, however, a difference between systems 'integration' and systems that simply 'interface'. Systems that communicate with each other in both directions may need to integrate, whilst systems that pass data in one direction only may only need to interface. So, what do we mean by the terms 'integrate' and 'interface'.

For the purposes of this chapter, we define integration as being where a system prepares data which are passed to another system that manipulates the data, adds value to them, and then sends the data back the first system. This description also applies to a number of systems which use common data files held in one place (e.g. as a database on a central system), such as names and addresses which may be updated from any of the different systems.

Interfacing of systems is where data are sent from a system in only one direction and either never, or very rarely, sent back again. This, for example, may apply to estimating data which are sent to a quantity surveying package once a tender has been successful. A simple way to describe the difference is that integrated systems need to communicate both ways, whilst interfaced systems tend to focus on one-off transfers.

7.2.5 Sequence

Another factor to be considered when implementing a multi-solution strategy is the natural sequence of events occurring in the business process. Consequently, you may wish to implement your systems in a structured order, for example:

- marketing before estimating;
- estimating before valuations; and
- planning before estimating.

This approach can be especially useful in, say, the case of valuation and estimating. To use a valuation system effectively a large amount of data needs to be entered and set up. Most of this information is derived from an estimating system. With an effective interface this information can be transferred and set up in hours or days rather than days or weeks. As a result, once surveyors are advised of the speed of the process it will encourage them to want to use the system, rather than making the excuse that it would take too long to set up. Another example would be the use of design briefing software before concept design software so that the information from one system can flow to the other.

In addition to determining a logical and practical business approach, there are other factors to be considered in deciding the implementation sequence. One of the most obvious is where there is a major cost saving to be gained by replacing a legacy system which is very expensive to run. The immediate savings gained through early implementation of a replacement system can help to pay towards the costs of the other new systems. Special needs – especially those of clients – may also change the preferred sequence of events.

7.2.6 How quickly is the solution required?

This will have a major influence on the selection of development methodologies. If a solution is required very quickly, with clearly defined deadlines, you may have no choice but to use packaged software. If, however, you have plenty of time then you may opt to develop your own software. The pros and cons of software packages versus bespoke software are discussed later.

7.2.7 Timescale and resources

Let us now consider the overall timescale of the plan. The 'selection to implementation' period needs to be as short as possible. Once you have started the process it is important to try and complete it as soon as possible. There is nothing worse than whetting people's appetites and then leaving them waiting for months on end with no apparent activity. To guard against this there are two things to consider here.

- Can the system be implemented in the period available?
- Do you have the resources to implement the system correctly within this time-frame?

Along with these considerations, you also need to manage expectations.

7.2.8 Expectations

Maintaining user expectation and keeping users informed is important, for long delays in systems implementation can create frustration and eventually loss of enthusiasm.

If you have been specifying and selecting systems for a significant period of time and have announced, for example, that they will be installed commencing month '*x*', user expectations will be high. However, if you have 12 elements to implement and then only complete them at a rate of one per month, it will be 12 months before the last unit gets into the system – and by then some of the key people who selected the system may have moved on. This is obviously not in everyone's best interests.

To reduce this period you will need to consider additional implementation resources. The business or unit into which you are implementing the system also needs to consider the question of its own resources.

Businesses very rarely have spare resources, and with IT you are usually replacing or updating a current system which at the same time must also be maintained to ensure that data remain relevant and usable. This is especially true when it comes to accounting systems. As a result it is not uncommon to parallel run an old system together with the new one, until such time as confidence is gained in the accuracy of the new system. However, running two systems is at least three times the work, so you may need to employ temporary resources to help with the implementation.

7.3 Software selection

7.3.1 Specification

Once the basic requirements for systems are known it is essential to develop a more detailed specification, commonly known as the 'URS' (user requirement specification). In achieving this it is important to determine and obtain details of all the requirements of the potential users, and then to categorise them using the following priority guidelines.

- Essential – the system must have these features or it cannot be used.
- Beneficial – the system can work without them but they will add value to the system.
- Nice to have – this is the icing on the cake. If the system can provide these features, all well and good, if not you can manage without them.

The process of developing the URS is more of an iterative process than linear. It starts by one person putting down on paper their requirements, then passing it to others who will be using the system and asking them to comment and add their own requirements. One person should be nominated to coordinate the responses and produce a regularly updated specification. The process continues until the final document is complete.

A good starting point for this process can be to review the appropriate packages on offer from various software suppliers as these will give users an insight into many of the typical features currently available. As mentioned earlier, in developing the user specification, it is important to refer to the business process results that you wish to achieve and the business strategy and IS/IT strategies that should be driving the implementation.

7.3.2 Software packages versus bespoke software

When determining how and from where you source your software, you may need to choose between off-the-shelf software packages or tailor-made (bespoke) software. In this context the important things to consider are the following.

- Development time/software availability.
- How long would it take to develop our own software?
- Are software packages available?
- Would such software packages need modifying?

To answer these questions it is worth examining the different approaches involved in considering both package software and bespoke.

Packaged software

Packaged software is the term used to describe software that is currently available and which can be purchased as a complete package from a software supplier. Software packages have usually been developed to cover a whole-industry requirement and therefore, by their nature, they will tend to provide a less than 100% fit for your exact purpose. If you compare your URS with the features of the package software, you may decide that, say, a 70% fit is acceptable, always providing:

- it meets all your essential requirements;
- what it does, it does well; and
- it is an improvement on the current system, for example, offering the ability to integrate with other systems.

A common problem with many software packages is that they meet most of your requirements but they need to be enhanced in some way. This is the way packages often develop from their original concept.

However, care should be taken with the amount of bespoke work you carry out. Wherever possible make sure that any changes undertaken are made to the base package and not just your version of it, otherwise your version may end up so far out of line that you are unable to take advantage of new releases of the original product.

One of the big advantages of software packages is the way they are constantly developed and kept up to date. The package gets the advantage of other users' input, not just your own. Consequently, future upgrades may add additional advantageous facilities that you hadn't even thought of. Overall, it should be said, packages usually receive better support.

Bespoke software

Bespoke software is that which is developed either in house by your own people or by others to your specific requirements. Such software should, therefore, provide a 100% fit to your needs, and this is the main advantage.

If seeking to use an external organisation, one alternative way of developing your bespoke software is to establish a partnership/relationship with the software supplier. In this case they develop the software to meet your particular needs, but they retain the right to sell your system to others. This is usually the case where your application – or parts of it – may have wider appeal in the general marketplace.

The development time for bespoke software should not be underestimated. It is not unusual for users to get the software specification wrong on the first couple of attempts, so it is better to use some type of rapid development prototypes to pave the way before you start on the coding details. This will indicate to users what they are getting and how their screens will look.

If you do not develop the software in partnership with a suitable software supplier, your software tends not to develop much further than its first form. This is because once you have got your basic software modules right, they tend to stay at the same level because of the lack of resources. Furthermore, unlike with general supplier-driven software development or packaged software, you don't get the feedback from other users.

Finally, do not underestimate the need to support any software that you have had developed. Even the best developed software will contain bugs, so that for the first couple of years your programs will need to be constantly modified to ensure that you meet the requirements of your users. In-house development therefore needs to be considered with care. Remember, it is probably not your core business to develop computer software.

7.3.3 Finding suppliers

Whether seeking standard or bespoke package solutions, you need to start by finding suitable suppliers. These can be located by reading the construction computing press or by consulting organisations such as CICA (Construction Industry Computing Association) who publish software catalogues in the UK. For bespoke software, particularly, you need to choose consultants/software developers who have satisfied clients, who have a good track record of success and who are experienced in developing similar types of systems to those that you need.

7.3.4 Selecting a package

Arrange for a small group of users to see initial demonstrations of what may be appropriate packages. This helps you to get a feel for the features currently available. Reduce a long list to a smaller one by eliminating those packages which do not fit into your technical environment or those that are too general. Try to bring this list down to about six possible suppliers.

Complete your URS with any features observed at the demonstrations and then send it to the potential suppliers, requesting them to provide specific demonstrations/workshops to enable them and you to go through your requirements in more detail. When selecting software that covers more than one discipline you may need to hold more than one demonstration.

Following the demonstrations let the users score the various features of each supplier's package against the URS and then select the most suitable product. At this point you will need to use the product with real data in a 'hands-on' environment - with your own users inputting and outputting from the system. This is often best done at the supplier's premises, because at this stage you may not yet have had sufficient training.

Once you are satisfied that the software will meet your requirements you should determine an implementation programme with the supplier. At this point you must also agree the level of resources that both you and the supplier will be putting into the project. Then you need to pilot the package.

Select one section of your company to use the software first as a pilot. During this initial period carefully monitor the implementation process and note any changes that you may require to be made to the software. However, do not make changes too early on in the project. Before making unnecessary changes, bear in mind that people often wish to carry on doing things the way they have always done them. As a general rule people are resistant to change, but that should not be the reason why good software needs to be altered. Consequently, investigate with the supplier and your staff what alternative methods there might be of using unmodified software in order to achieve the desired features that users want.

Identify any essential changes required and agree with the

supplier the method of achieving your goal as originally stated. Unless software changes give your company a significant competitive advantage, always ensure that it's the base product that the supplier agrees to change so that your version is the same as the rest.

Software development

When it comes to bespoke software, agree the specification with the users and use as many diagrams as possible – showing them the data flows, screen layouts, etc. Where possible use rapid prototyping methods. Hold workshops between the users, designers and coders to discuss the specification, always ensuring that the software designer understands exactly what you are trying to achieve.

It's important to agree a programme for the work with the software developer. This should be broken down into manageable sections and should include milestones for deliveries. Next get a selected number of users involved, whose duty will be to review the progress and approve the delivered sections of the software. Then agree who will 'sign off' the system.

Also determine what exactly are the change control procedures that are to be used, ensuring that everyone involved, especially the project sponsor, is well aware of any proposed changes to the specification. Then, as with package software, the next stage is to run a pilot. Choose a division of the company, section or project which has the most interest in the system and monitor their use of and reactions to it.

Carefully document the problems they encounter and any suggested changes to the system. It is important to ensure that problems are fixed quickly, but, if you can, avoid making changes until you have finished a reasonable amount of testing and use. Then review the proposed changes with the project team and agree the changes required. This process may need to be repeated a number of times depending on how good the original specification was.

7.4 Implementation

7.4.1 Skills audit

By this stage you should have an agreed solution, so the next step is to ensure that users have the basic skills which enable them to take advantage of it. To determine this, carry out a skills audit to gauge the level of expertise in the section where you intend to implement the new solution. This is especially important when you are changing from one type of user IT environment to another. A typical example of this would be moving from a dumb screen application to a windows environment.

Remember, whilst some of your users might be experienced in using an existing package, they might not necessarily know the first thing about computers of different types. This audit stage can be carried out in parallel with any testing or piloting.

7.4.2 Data preparation

When it comes to data preparation, you need to decide what data are to be transferred from any existing system to the new solution. The time to clean up any old data is now. Once you have transferred the data they will not get cleaned again, as everybody will be too busy to find the time.

Determine the best methods of transferring the data from one system to another with the minimum of manual intervention. Also make sure that you validate any data you transfer. It is sometimes only too easy to electronically enter data into a database, only to find that, when a user accesses that field at some later date, they may not be able to leave it because it contains invalid data. An example of this would be names and addresses that have been transferred, but where the county is erroneously in the town field and there is no postcode. If these fields are validated then the system will not allow you to leave until you correct them. If a user only accesses a record for information and finds they cannot leave it without providing the correct information they can get very annoyed with the system.

7.4.3 Training

Training is the key to any successful implementation, but it is important to ensure that you have the right material. Train personnel on relevant data and in the way that your company wishes to use the software. This is especially true when using a complex package that can be used in many different ways. Also tailor the training to suit the users. At the same time it is useful to identify super-users, individuals to whom other people in the department can turn when they are unsure of an aspect of the software and its use.

When implementing a complex or integrated strategy, it is important to agree, with both users and suppliers, any phasing that may be required. This must always take into consideration the degree of resources that are required to carry out the implementation, both at the user level and implementor level. Carry out the training as close to the implementation as possible and also ensure that users can carry on using the software when they return to their workplace. There is nothing worse for users than to receive training and then not be able to practise what they have been taught. People soon forget what they have been taught, and when they do you will then be involved in retraining them.

7.4.4 Revisit users

Once the solutions have been installed you will need to revisit the users to ascertain how well the solution is working and whether they are using it in the right way. On a multi-solution implementation, progressive release of the software is advisable, as the users become familiar with the relevant parts of it.

Take care when implementing any upgrades or patches to live systems. Document any changes that the upgrade or patch might involve and circulate to the users, telling them when the changes will take place and what those changes are.

7.5 Support

7.5.1 Options

Implementation strategies do not end simply with the implementation of the software. Any system needs to be supported, and you will need to consider how the help will be provided to support the users. The first question is: 'Do you use the software supplier ?' This is often the usual way, but there are other options.

7.5.2 Help desk

You could consider setting up your own help desk to provide a first line of support, filtering basic enquiries and resolving easy problems, before you need to use the supplier's external support unit. Support fees can sometimes be reduced if you are able to filter calls first. One important aspect of establishing help desks is that it improves the reporting of faults or problems and the analysis of these. Reviewing these regularly may point to the need for additional training or even a change in the system to avoid recurring problems.

7.5.3 User groups

A good way to maintain the development of a product is through user groups. These can be internal or external. An internal group helps your users to develop their skills in using the product, while external user groups are good in encouraging the further development of the product itself. In either case it is important that these groups are led by the users. It is one sure way to maintain an IT focus, involving those for whom IT development is usually really meant.

8 The way forward

Martin Betts and George Ofori

"By the next millennium, IT will have changed our business to allow us to always quote fixed prices."
– Managing director, UK civil engineering contractor

The current use of IT for strategic purposes in construction is rather limited. There are a number of strategic management techniques, IT planning techniques and principles of strategy implementation available, but how these get taken forward by construction companies is difficult. The way to overcome these difficulties at different levels in construction embraces exploiting new opportunities, developing an integrated strategic framework, systematically applying a range of techniques at different levels, and managing the process of taking this forward.

- Taking forward more effective strategic exploitation of IT in construction needs further work by national agencies, professional institutions and trade associations, companies in isolation and collaboratively and innovators amongst project teams.
- In seeking more effective use of IT in construction, the questions that need to be addressed are: who should do it, what should they do, how should they do it and why.
- The various strategic management and strategic IT planning techniques all have potential application in the various operating levels of the construction sector.

"Information should flow from the building site to general management and then back to the site."
– French project manager

8.1 Introduction

This chapter draws together the different strands of ideas presented in this first part of the book and projects them into the future. We see how the issues of strategic IT planning and business process analysis and of SISP and strategy implementation may come to be applied throughout the construction sector in the future.

8.2 A five-level framework

8.2.1 Further opportunities for strategic application at the national level

The Singapore and UK experiences show how construction in a country can gain considerable benefits from a coordinated national approach to its IT use. Other countries can similarly look to follow such an approach. Evidence of similar developments in Denmark, Finland, Sweden, South Africa, France and Australia is starting to emerge.

8.2.2 Further opportunities for strategic application at the professional level

As our examples have shown, certain institutions within the construction industries of some countries have accepted the need to think strategically and in a process-based way about their use of IT. The next task is to convince them to do this in a structured manner. The concepts of strategic IT planning formulated by Porter and Earl and of business process analysis by Davenport would be very useful in this regard. The different SISP techniques would also have their place at the national level. It is also necessary to persuade other institutions of the need to act. How can this be done?

- Sufficiently large institutions can undertake studies with in-house expertise. Another possibility is to appoint specialist consultants.
- Institutions need to identify specific research, development and procedural change programmes that are necessary. These should be set out systematically, perhaps after further consideration by a study group or an external consultant.

- The resulting set of mission, objectives and action plans should then be communicated to members. Ideally, the institution's strategic plan should provide the basis for the enterprise level planning of its members.
- Guidelines on how members could adopt or implement the contents of the plan should also be prepared, and well publicised.
- Rather than continually preparing new strategic plans, it would be beneficial for each institution to ensure that what they draw up is actually implemented. A set of procedures for monitoring the extent to which its members have acted upon the stated initiatives and the results achieved would be useful.

Like plans for enterprises, those at institutional level should be of a rolling nature. The monitoring of their implementation would facilitate any necessary review. Consequences at the institutional level are at their most significant. Failure to act in the current climate may mean that changes occur in a profession's relative position and strength, from which recovery is impossible. Grasping strategic opportunities earlier than other professions could, conversely, result in a leading position in the construction sector of the future. Institutional objectives, therefore, should be survival and inter-professional advantage.

Some professional institutions and trade associations in the construction industries of many countries, like some of those described in this chapter, have applied Porter's (1990) diamond planning concepts but there are many exceptions to this. Despite the current dynamic business climate, there are old, traditional and well-established statutory protected bodies that are currently well-recognised and regarded. Architects and engineers are good examples of these in some countries. It could be argued that these bodies need to follow strategic management and process analysis much more earnestly than those which have responded from the disadvantaged and uncompetitive positions in which they had found themselves. The dangers of complacency in the current dynamic business environments are greater. Experiences of individual, uncompetitively positioned enterprises, and groups of enterprises in uncompetitively positioned national industry segments demonstrate this. There is no reason to suggest that institutions in construction will escape these influences. Yet there have been enterprises that have exploited the changing environment positively. Construction professions could do the same.

There is great potential for inter-professional strategic management and process analysis and synergistic developments leading to mutual benefit. Professional institutions and trade associations must continue to explore these and other issues in their strategic management and process analysis in the future. As the structure of the sector changes, there will be scope, in each country, for strategic alliances among all, or groups of professional institutions and trade associations. Such alliances and/or cooperation could be for addressing particular issues, or relate to wider matters affecting construction. Such multi-professional, and preferably industry-wide cooperation and alliances are most relevant in the developing countries where the ability to influence government to adopt favourable macro-level policies and procedures is vital to the development of the local industry.

Before interests become too fragmented, efforts should be made to fundamentally restructure the professions and create institutions which are more appropriate to the needs and circumstances of the country. Linkages across national borders are also potentially beneficial, especially with the lowering of national trade barriers. Such linkages may be between or among institutions of the same profession, as well as those of different professions. For example, a strategic restructuring of the professions in the UK undertaken together by all the relevant institutions (rather than the present uni-professional approach) to bring them more in line with the structure of the industry in Europe is a viable option. At the same time, the opportunity could be seized to restructure all professions world-wide to address the problems they are likely to face in the future. Greater cooperation among the professional institutions and trade associations not only on a continental or regional basis, but also on a global one, is necessary. IT is the key to such developments.

This chapter has shown that strategic management and process analysis, in response to dynamic factors within their industries and their operating environments, will affect the future of individual professional practitioners in two ways:

- by helping their professional enterprise to explore the opportunities it has for competitive advantage over other enterprises; and
- by enabling their professional institution to explore professional level competitive advantages.

The net result will be a more competitive environment within the construction industry. Construction practitioners must take due cognisance of efforts by professional institutions and/or trade associations to ensure their survival and enhance competitive advantage.

8.2.3 Further opportunities at the enterprise level

The techniques of strategic management and process analysis are not mutually exclusive options which should be tried one at a time. Moreover, their purpose is to enhance the firm's adaptability to change rather than offer immunity from the implications of change. For the construction enterprise, formulating a strategy should not be approached from the same viewpoint as a lottery. Nor is it a matter of coming up with something new every now and then. The important issue is to develop a long-term view based on the company's core competencies, which should be continuously improved over time to offer the necessary competitive advantages and seek new niches to exploit.

Strategic management and process analysis are relevant to all construction enterprises, large or small, regardless of the aspect of construction they are involved in. They should be recognised as an important aspect of the firm's overall activity which requires as much attention as its routine operations. The strategic management process should be approached in a structured and systematic manner. The increasing body of work, not only on the techniques of strategic management and process analysis, but also relevant corporate approaches to their direct application, should be studied continually by construction enterprises to identify potentially beneficial elements.

The responsibility for strategic management and process analysis should be specifically assigned to a person or group within the organisation. CSC Index (1994), as a major BPR consulting firm, advocates that, without the support of senior management, BPR is unlikely to be successful. To this end, the variability of enterprise size must be recognised. For large international constructors, there is a clear and growing need for strategic management and process analysis teams upon whom the long-term survival of the enterprise may depend. Many larger enterprises may already have such teams. Where they do not yet exist, the task of strategic management and process analysis may be assigned to the 'business development' division within the firm. These divisions should be staffed by multi-disciplinary teams, incorporating construction professionals as well as specialists in corporate strategic management and process analysis. They must be placed and perceived as central to current and future company plans rather than as peripheral service departments. The departments should apply strategic management and process analysis concepts in detail to the specifics of their current position and the resources of the company, giving due regard to corporate objectives. Many innovative strategic initiatives will emerge in the construction industries of many countries in the immediate future if such departments become widespread. The consequences of this will be restructuring, the creation of new businesses and the emergence of new products and services.

There are many small and medium-sized construction enterprises for which strategic management and process analysis teams may not be sustainable. This does not mean that the issues are irrelevant to such firms. Articulating their vision for the company and formulating a set of objectives and a plan for achieving them would be a useful exercise for the proprietors of such firms. For owners of firms in this group who are unable to do this by themselves, consultants may be engaged. Small firms in developing countries would benefit from central assistance (by a contractor-development or management-development agency, or a committee set up or consultants engaged by the contractors' association) to formulate corporate strategies using the emerging techniques.

As we have seen earlier, enterprises outside the traditional construction industry (such as materials manufacturers) may consider construction activity within their strategic sights as they seek to vertically integrate or diversify their activities, or generally find new business. Moreover, as deregulation strips the construction professions of the protection offered by legislation in some countries, the trend of firms and persons from outside the industry (such as accountants and management consultants) offering construction-related services will be accelerated. The need for construction enterprises of all types to view strategic management and process analysis as a vital activity cannot be overemphasised.

The strategic management and business process initiatives of construction enterprises can be stimulated by, and benefit from the efforts of public agencies, professional bodies and trade associations (such efforts would also be helped by the application of the techniques and tools discussed above).

In Part A of this book we have reviewed the application of strategic management and business process analysis by enterprises in the construction sector. We have seen that the business and construction context is increasingly dynamic. The new concepts of competitive advantage, and techniques by which it can be achieved (advanced by Porter and others), are shown to have direct application in construction.

There has been much previous work by researchers into strategic management in construction although process analysis has not attracted as much attention due to the relative recent emergence of the way of thinking. Moreover, Porter's, Earl's and Davenport's concepts have many parallels with the way that others perceive the problems and issues. However, the need is currently more pressing, as the Egan Task Force report (1998) and QS Think Tank report (1998) demonstrate. There is scope for improvement in practices in construction enterprises. The emerging concepts of strategic business management have been successfully applied by leading construction enterprises to gain competitive advantage. Even more than before, construction is now a highly dynamic sector whose operating environment, industry structures and product requirements are changing at an increasing pace. These changes bring an increasing requirement for construction enterprises to exercise more sophisticated and systematic strategic corporate management. Despite the particular hindrances to the application of the existing strategic management and process analysis techniques for construction enterprises, there is growing evidence that the leading edge of current practice throughout the world is beginning to act more strategically in its business planning.

All construction enterprises will ultimately have to consider strategic concepts to be able to operate effectively in the emerging industry context. In doing so, the techniques presented here will need to be applied, whilst recognising the hindrances presented by the nature of construction, and finding ways of overcoming or minimising them.

There will also be opportunities at the project and product level but these are mainly actionable by enterprises in a similar way to that described here. Beyond the wider adoption of strategic management and process analysis, it is also clear from this part of the book that IT has considerable strategic potential at a number of levels in construction. Nations, institutions and enterprises have a range of SITP and SISP techniques and models available to them with which to enhance their IT exploitation. Well-established practical guidelines for IT implementation are then available to realise these strategies.

8.3 An integrated framework for construction

The recurring considerations made in this chapter about better strategic use of IT have focused on our quest for integration of information. Current fragmentation and other constraints in the sector, and the realisation of BPR, largely require some improvement of information integration. This prompts us to ask several questions regarding the subject. We started to wonder if integration of information is our holy grail, what do we mean by it? In particular:

- who integrates what information?
- how and when an organisation should integrate information?
- why an organisation would choose to integrate information?

This led to the development of the framework presented in Table 8.1. This framework can be used to define dimensions and levels of integration in construction information. Many organisations looking to achieve greater strategic exploitation of IT are likely to need to achieve greater information integration. This is a very open and diverse goal. This framework may help organisations plotting

Table 8.1 Dimensions and levels of integration

	1 Low integration	2	3	4	5 High integration
Who?	Individuals	Departments	Entire organisations	Whole project life cycle	Entire industry
What?	Data	Models	Knowledge	Goals	All project information
How?	Islands of automation	Multiple applications in one discipline and phase	Multiple applications from several disciplines in one phase	Multiple applications from several disciplines and phases	All applications in the project delivery process
Why?	Survival	Increase profit	Increase market share	Enter new market	Create new market

their way forward to better understand their information integration motivation, action plan and objectives.

With regard to who, we can imagine integration among individuals and departments leading to the integration of entire firms and projects and ultimately the entire AEC industry. To answer the question of what to integrate, as an initial step we might choose to focus on sharing just data. This could then be expanded to include models, such as product and process models, knowledge about decisions, and project goals. Ultimately, data, models, knowledge and goals would all be shared.

With regard to how we integrate, just a few applications within one phase and discipline might be a starting point which then could be expanded to include all applications from all disciplines and phases.

Reasons why anyone might integrate or increase the level of integration are to:

- stay in business;
- increase profit, market share and market size; and
- enter or even create new markets.

That information integration offers these opportunities has already been demonstrated by such companies as OTIS (Cash and McFarlan, 1990) along with many others from outside of construction. In Table 8.1, we attempt to list the various values of these four dimensions of integration to indicate increasing levels. However, it is not always necessary to tackle the steps in order. For example, a firm could well opt first to integrate goals and then tackle knowledge or model integration. Yet a framework such as this is important to distinguish different forms and stages of integration and for all in practice and research to be able to cross-refer to different initiatives using a common understanding. Earl (1989) also argues that such a temporal framework is important in identifying sequential stages of progression in both research and development and practice.

The framework allows individuals, departments, companies, projects and industries to plot their current state of integration and to indicate efforts to increase its level. Thus, the framework becomes a vehicle for comparison and for focusing development and implementation efforts. For example, two departments might differ in their capabilities of sharing project information, or a company might be interested in pushing its integration capabilities from level 3 to level 4 for the how and when dimension.

This framework also provides a generic and focused definition of integration. Generically, integration can be defined as the sharing of something by somebody using some approach for some purpose. Obviously this is not a very useful definition. However if one substitutes the vague expressions with values from the framework, one can create a definition that suits a particular purpose. For example, a firm might define integration as:

"the sharing of data and models by departments using several applications pertaining to a number of disciplines and project phases for the purpose of increasing profit and market share."

Another company might define integration differently because it has a different purpose or image. The advantage of the framework is that different definitions can be related to each other easily. The framework has five dimensions that are each presented in Table 8.1 independently. In reality, each dimension can be combined with others in a multi-dimensional way which we can illustrate by combining the dimensions why and who as in Table 8.2.

8.4 The way forward

In Part A of the book, we have seen a range of criteria by which IT use can be considered strategic. The examples from the five levels of our framework show a number of areas in which the construction sector is being strategic in its IT use by virtue of the concepts within our SITP and SISP techniques being adopted. Our answer to the questions posed here, on the basis of the review performed, is that we are attempting to use IT strategically in construction in a number of ways. However, the picture is a patchy one and represents the leading edge of current world practice rather than industry norms. Table 8.3 summarises the present situation by showing strategies that appear to have current and potential use at each level. These are evaluated from the examples of current IT applications in construction which we have reviewed above.

What Table 8.3 represents is a summary of much of the material discussed to date. It is a matrix of the five application levels and some of the different strategy and process concepts and technology we have looked at. Each cell in the matrix represents a potential application of a technique at a level. The number in each cell represents our assessment of what is happening. A number 2 in a cell shows a combination that directly arises in our examples. A

Table 8.2 An example of the interplay of two dimensions within the framework

	Who?				
	Individuals	Departments	Enterprises	Projects	Industry
Why?	Stay in a job	Department survival	Stay in business	Complete project	Industry survival
	Increase earnings	Increase profit contribution	Increase profit	Increase project success	Increase industry profitability
	Extend job authority	Increase department political role	Increase market share	Increase project scope	Increase industry share of economy
	Change jobs	Assume additional roles	Enter new market	Make project more widely useful	Extend into other sectors
	Create a new job	Create new roles	Create new market	Create project extension	Create new sectors and economic services

number 1 shows combinations that appear likely. The cells without numbers may also offer strategic opportunities. The use of a matrix is instrumental in showing possibilities and identifying where they are most likely to be found.

We have also seen some examples of IT being used for core competencies and core capabilities strategies. These have shown the need for the whole construction sector to examine the potential for strategic use of IT for competitive advantage. We are being strategic in our use of IT in construction, but not sufficiently so. There needs to be a change in our approach in IT research and practice in construction. We must move away from attempting to use the latest technological advances to automate parts of our current manual processes towards being creative in exploiting the potential IT offers for:

- new products and services;
- new businesses;

Table 8.3 The application of strategic and process methodologies to the applications framework

Part of strategic approach	Application level				
	National	Professional	Enterprise	Project	Building
Value chains					
Lower cost	1	2	2	2	2
Higher value		2	2	2	2
Value channels	1	1	1	2	1
Five forces model					
New entrants		2	1		
Supplier power		1	2		
Buyer power		1	1	2	
Substitute products		1	1		1
Jockeying for position		1	1		
Generic competitive strategies					
Product differentiation	2	1	2	1	2
Cost leadership	1	2	2	1	2
Product focus	1	1	1	1	2
Business process analysis	1	1	2	2	

1 Obvious opportunities for strategic use of IT.
2 Areas where IT is being used strategically in construction at present.

- different processes and procedures; and
- project and industry structures.

The opportunity for a 'big bang' in construction exists. Our problem is not a lack of technology but more a lack of awareness of how to exploit it and of how important major process and culture change is to allow this to happen.

There are a number of preconditions that must be met before more purposeful exploitation will happen. First, there is a need for realisation by practitioners operating at all levels within the construction sector of the need for strategic approaches to be taken in our IT development and planning. Second, the techniques and tools by which strategic applications and process analysis can be identified must be examined and analysed further. The techniques are not simple tools that can be taken off-the-shelf and applied easily in construction to achieve dramatic strategic implementations. There are many hindrances that are presented by virtue of the nature of the construction sector. Similarly the current trends and developments in construction give rise to possibly unique strategic opportunities for us. We must therefore explore further the way that strategic IT applications can be made specifically by us in the construction sector.

Part A of the book has defined a classification of five application levels at which strategic IT planning can be used. It provides researchers and practitioners with a broad picture of where further development studies are required and of where priorities lie. It should draw attention away from the concern with automating existing processes and toward the need to target strategically.

For construction industry practice, the benefits are in awareness of how the use of IT in the sector is going to change. This will have fundamental effects on the way the industry works. The challenge is to ensure that the ideas are applied in the manner suggested in Chapter 7 on implementation. This will be the case for all practitioners. This is illustrated further in Chapters 11, 15 and 16. We have shown evidence that, if this is done, commercial benefits will accrue. We return to this issue in more detail in Chapter 12.

A new IT research and development agenda for construction practice emerges from this analysis. Individual studies within unique organisational settings are now required to examine strategic opportunities for IT exploitation at our five application levels. The way that these studies should be carried out will eventually depend upon the particular requirements and circumstances of the nation, profession, enterprise, project or product concerned. An indication of how this may be achieved and strategies appropriate for each is now available for the first time. This issue is developed further in Chapters 13 and 14.

8.5 Summary

We have seen in Part A of the book that the concepts of strategic management and process analysis are widely applicable in construction and that IT has a central role to play. In practice, these concepts are applied less widely in construction than in other industries. However, construction is beginning to embrace these issues more widely. In five years time, we expect these concepts to be widespread in our industry and commonplace within ten years. Beyond that, it is likely that the competitive nature of construction will have changed considerably. Whether this will be to the benefit of you and your organisation may be partly determined by your reaction and that of your colleagues to these concepts and the way you apply them in specific situations.

Part B Current IT Practice in Construction Businesses

Part B of the book is the smallest of the three. It builds from the introduction of new ideas and thinking from Part A by explaining and describing what current practice is in using IT in construction businesses. This is done in overview form initially by describing the general principles that emerge from 13 major case studies of IT management from the UK and elsewhere in the world. This is followed by presentation and consideration of results from detailed best-practice benchmarks that have been undertaken of IT use in support of business processes.

Benchmarking, as a technique, is revisited in detail, following the introduction that has been made to the subject in Part A. The results of benchmarks of IT support to discrete construction processes are then described, looking at all parts of the construction project life cycle. Typical construction IT practice for each process is described and contrasted with world-class best-practice use of IT from a range of other industry sectors.

The part concludes with a description of results from benchmarks of how IT is used to achieve project information integration from six major construction projects from throughout Europe. As with other parts of this book, all chapters are illustrated with managers' quotations, case studies and one-minute manager summary pages.

9 Current strategic practice

Martin Betts, Martin Jarrett and Mathew Shafaghi

> "Projects are now so complex, cost control and project monitoring can only be done through IT."
> *– Regional director, German contractor*

The process of strategic exploitation of IT in construction is influenced by a number of key competition and business strategy drivers. The process is made possible through a number of mechanisms of effective IT management. There are 'nuggets' of good practice in the way IT is used to achieve competitive advantage in current construction. They are often specific to particular companies and their contexts and environments. Opportunities for strategic IT exploitation are often triggered by regulatory change, political and economic events and major social issues.

- Nuggets of IT exploitation are to be found in specialist foundation work, industrial relations, purchasing activities, health and safety, design processes, tracking business leads, managing cashflow and marketing.
- The main drivers of strategic IT exploitation in practice at present are a company's core competencies, their competitive behaviour, the formation of strategic alliances and different visions for the role of IT.
- Enabling mechanisms for effective IT management include innovation activities, the management of IT departments, users' involvement, strategy development and implementation processes and IT appraisal methods.

> "Projects are managed on the construction site."
> *– Italian project manager*

9.1 Introduction

The objective of this chapter is to analyse the current strategic role of IT and the way it is used for business efficiency and competitive advantage in construction organisations at present. In doing so, the chapter seeks to answer the following questions.

- How are construction companies benefiting from the advances in IT?
- How do IT strategies assist companies to maintain and gain competitive advantages?
- Is there a simple recipe being used for developing effective IT strategies?
- Is there a formula for successful implementation of IT strategies?

In compiling this chapter, ten of the UK's leading construction organisations and three international construction organisations were studied as detailed cases to find out how IT is managed. In addition, 13 examples where IT was exploited by these companies for strategic benefit, in various processes of construction, were documented and analysed.

To gather evidence from a wide cross-section of decision-makers and IT users in the organisations studied, qualitative techniques of data collection were used. Over 150 managing directors, IT directors, IT managers, financial directors, commercial directors, business development managers, design directors, marketing directors/managers, operational directors together with a number of senior, middle ranking and operational managers and personnel from construction organisations were targeted for structured, semi-structured and group interviews, generating over 2000 pages of information about current practices.

The collected information from the above led to the development of patterns and propositions which were shared by all of the companies. Some of these propositions are shown in this chapter. The propositions provided the basis for the development of a model of the issues involved in effectively planning to strategically exploit IT in construction; this model is represented in Figure 9.1.

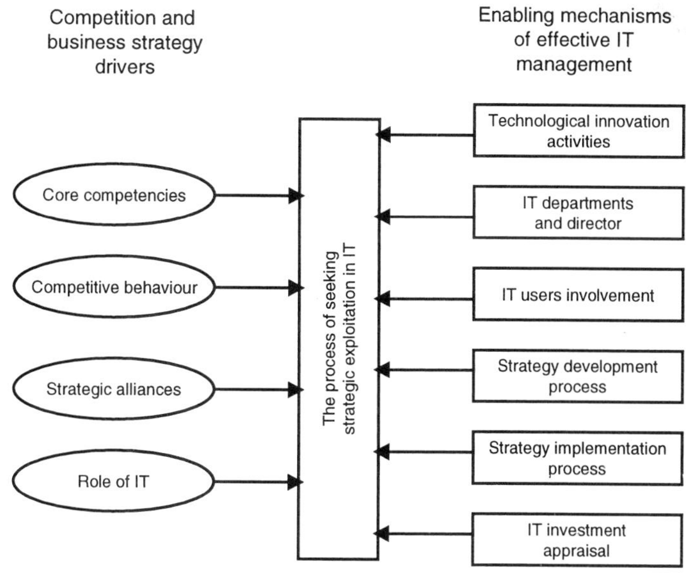

Figure 9.1 The strategic IT exploitation process

9.2 Data collected

The next sections of this chapter are illustrated with pieces of current evidence from industry. These include:

- brief summaries of industrial cases; and
- propositions developed from study of the cases.

They are presented as alternative sources of information to suit the needs of different audiences for this work.

9.3 Problems in the construction industry

As we have seen in earlier chapters, the construction industry world-wide stands on the threshold of fundamental changes. These changes are causing all construction organisations to radically review what business they are in, what products and services they provide and how to be more competitive. A principal feature of how companies in other industries have coped with these changes has been to strategically exploit IT. Construction now has the opportunity to do the same. There are some examples of leading construction organisations who are already responding to this need for change through IT. All construction organisations must now do so if they are to continue to compete in future. This chapter sets out to inform the process by which this can be done.

Construction companies believe that IT is critical to the survival and prosperity of their organisation

We are all aware that the construction industry is a mature and traditional sector that is slow to change. The industry knows it needs to fundamentally break out from its traditional ways of doing things. There is a need for those involved in IT management to arm themselves with a more ambitious vision of how IT can fundamentally change the way that the industry works and to strive for it. We can paint a future scenario for construction of where:

- IT provides a major integrating force for the problems of fragmentation to be overcome;
- companies have more accurate and timely information about their projects and their business;
- IT enables new and improved products and specialist services to be provided;
- quality and certainty of outcome are improved; and
- the value added to clients from an improved construction process is demonstrable.

These are the opportunities and the vision towards which all construction organisations should strive. They should seek to do so partly by strategically exploiting IT.

9.4 Competition and business strategy

9.4.1 Core competencies

IT expertise in some companies has led to the development of core competencies

We have found that sustainable profit is sought by construction companies through deepening and enriching their distinctive culture through their core competencies. Core competencies are seen to allow construction companies to focus on innovative and profit-making activities in line with the company's corporate goals and objectives. Industry also indicates that much of the success in this area is due to the need to serve new markets and achieve technological advances and operational excellence, superior design, engineering solutions and an improved construction process.

Experience from industry suggests the following.

- IT supports companies' core competencies.
- IT is a critical factor in the development of core competencies.

- Construction organisations are increasingly benefiting from the competitive opportunities provided by their core competencies.
- Core competencies are difficult to sustain therefore construction organisations should transform them into income-generating and profit-making activities.
- Construction companies should divert part of their revenue from core competencies towards R&D with a view to maintaining their competitive edge.
- Core competencies are often developed at the divisional or business unit level.

Case study: Global positioning of piling rigs

The system is the only GPS-based real-time positioning system of its kind in use today. Designed specifically for hydraulically operated piling rigs, the system allows a driver to accurately position the rig over a pile position. Working to centimetre level accuracy, the system gives a reliable, accurate and innovative alternative to setting out. This technical core competence is being exploited by the company in gaining a competitive advantage through differentiation in niche markets.

9.4.2 Competitive behaviour

It is clear that the profitability of construction companies is determined by their competitive behaviour, through innovation and the provision of products and services of value to their clients. One director told us that, 'Being innovative helps to keep us ahead of our competitors.' Another managing director commented, 'Our one-stop-shop approach has given us a competitive advantage over our competitors.'

Experience from industry suggests the following.

- Company structure, systems and people play a critical role in the success of a construction company's competitive behaviour and strategy.
- Construction organisations should focus on how best to serve clients and reduce costs through process improvements and technological innovations.
- IT can increase the pace at which the company can evolve to meet new demands that market changes call for.

- Construction companies should use IT for design and implementation of competitive strategies.

9.4.3 Strategic alliances

Experience from industry suggests the following.

- To grow and secure market status, construction organisations should combine resources with the discipline of a large firm and the realisation of synergy between different divisions of a company as well as carefully matched outside organisations. They need a vision of the firm as a whole and where it can go.
- Strategic alliances are important as a competitive weapon and need to be based on long-term orientation of the organisations concerned, common objectives and trust.

In return, the rewards are high. We have documented five strategic alliances between construction companies and software houses that resulted in competitive advantages and two examples of strategic alliances of construction companies with specialised engineering and software companies that have led to market leadership and competitive advantages.

Industry suggests that construction organisations favour the IT strategies that can be developed at divisional level rather than integrated through the company as a whole. These strategies often encourage and foster competitive behaviour through internal and external strategic alliances. Although the extent and nature of collaborative ventures vary, construction companies have substantially benefited from such collaboration.

- **International collaboration as a part of global strategy**. We found three examples where IT was a critical success factor for establishing and managing international collaborative ventures.
- **Academic collaboration**. We found that in six of the cases academic collaboration had a significant impact on the development of IT initiatives.
- **Technology collaboration**. We found two examples of collaboration between construction companies and specialist engineering companies. In both cases, the strategic alliances have led to market leadership and competitive advantages.
- **Competitors' alliances**. Many companies are aware of the need and opportunities provided by these alliances.

- **Collaboration with the IT industry**. In five cases strategic alliances between construction companies and software houses led to perceived competitive advantages.
- **Collaboration with clients and suppliers**. We did not find evidence of collaboration with major components and materials suppliers. However, collaboration with clients was viewed by the majority of the companies as favourable.
- **User participation**. In all the cases there were some elements of user participation. But these alliances are limited and at the early stages of development.

Case study: Corporate purchasing system

Developed by a major European contractor, this purchasing system is capable of aggregating data across all projects and all lines of business. The aggregated data add value to the company by providing strategic information to contribute to negotiations with suppliers and subcontractors and allowing long-term relationships to develop with them. This use of IT exploits the strategic advantage of significant buyer power.

9.4.4 The role of IT

Some clients prefer to conduct business with construction companies that are leading the industry in terms of IT

Some construction companies believe they have gained competitive advantages through IT

Industry revealed that in the late 1990s, construction organisations are increasingly aware of IT and its value-adding potential and are appreciating the strategic opportunities it provides. Apart from the support functions, we found that construction companies are extensively using IT for process improvement, innovative approaches to engineering and construction problems, client satisfaction, management of competitors, and new contract and procurement systems. Therefore it is difficult to divorce the role and importance of IT in recent developments in the construction industry from its impact on the survival and growth of these companies. Although the lack of progress of construction companies in adopting IT is frequently blamed on the nature of the industry and the lack of investment in this area, industry suggests that the main problem lies in the way in which the company and industry culture allows IT to be managed.

9.5 IT strategy

9.5.1 Technological innovation

Industry suggests that the uneasy relationship between senior construction professionals and IT is due to a perception by management of a historical failure in delivering the expected benefits, excessive development and implementation costs, inappropriate IT facilities and process changes. To overcome these barriers, construction companies have looked ahead and developed IT strategies that are in line with their company's business strategy with consideration to current technological trends regarding IT. We found two companies that attempted to benefit from the proven IT technologies developed by others, four companies who used their own existing IT more effectively through a well-thought out strategy for achieving advantages, and eight construction organisations who sought to be in the forefront by developing and innovating with newly emerging technologies. From the latter group, three companies perceive that they have achieved substantial financial gains, market status and competitive advantage through being innovative first-movers in new technologies.

It is difficult to forecast innovations in the field of IT and the desire of some construction organisations to be at the leading edge of technology is risky because it is often accompanied with cost implications. Equally, falling behind is almost as dangerous. Hence, the most suitable strategy may be to focus on the utilisation of recently developed and proven information technologies that could lead to efficiency gains and competitive advantages. Links with leading research organisations were seen as being key to this. We will see in later chapters in Part C how emerging technologies have great innovative potential for construction.

Case study: 3D modelling

The 3D computer modelling system is an integrated project database incorporating graphical and textual data for use during the design phase of building projects. The system is concerned with how the whole life cycle links with planning and cost modelling. It also seeks to allow a streamlined construction process and a niche strategy on specialist projects such as in leisure, health and highways projects.

9.5.2 Construction and the IT industry

Software manufacturers, as a part of the IT industry, have benefited from substantial growth in recent years. However, evidence from our case studies suggests that their contribution to the construction industry leaves considerable room for improvement. We found that seven construction organisations have significantly benefited from the technological advances in the field of IT through strategic alliances with a limited number of software houses. In addition they have managed, through their bargaining power, to demand a more constructive and effective role from software providers. This new role includes promotion of standardised and dedicated software with integration capabilities, improved quality of service and maintenance and effective training programmes. Some companies have gone further by involving the software companies in the in-house development of their IT applications. Their role in this case is to act as a facilitator and to ensure full compatibility with the existing applications as a systems integrator. These benefits have been achieved through common goals, trust and good long-term relationships.

Case study: Health and safety

In this case, IT has assisted a construction company in the adaptation of the business unit/group safety policy to comply with government legislation (Health and Safety at Work Act) to ensure that as much attention is paid to safety of operators as to other factors of production. It has been successfully applied to offshore projects and has helped the company develop a differentiating competence in this area.

9.5.3 IT departments and directors

We found that nine companies managed IT through a decentralised and four through a centralised structure. Industry is of the view that a centralised structure is good for culture, vision, decision making and control, whilst a decentralised structure is good for support, user satisfaction and ownership. However, the role and structure of IT departments are changing in order to facilitate the exploitation of IT for strategic purposes. We found evidence that there are more senior personnel, thinkers and facilitators and fewer technical support staff. IT departments are playing a more active role in creating an IT culture among senior managers and improving the level of IT skills within the company. This is also reflected in the role of IT managers and directors. Industry suggests that whilst they are responsible for the overall aspects and management of IT resources, facilities, selling IT and providing guidelines and strategic direction for all the business units, they are also horizon watchers for new developments in the field of IT. Industry also revealed that IT professionals consider that their contribution to business strategy formulation would be important, as it would lead to the development of IT strategies that would most support the business activities and corporate goals.

IT directors/managers do not participate in the formulation of business strategy

We found evidence that, increasingly, IT departments are being directed by non-IT construction professionals with technology support staff working alongside them. There is a tendency for more of these directors to be represented at a more senior level within the management of the business. We have found evidence to suggest that both of these changes are necessary for construction organisations to more effectively exploit IT for their business.

Case study: Civil engineering CAD

This CAD system is used to improve the creation of geographical information and the flow of information from design to construction. The system is principally used for motorways but it is very useful for any project involving earth moving. As such it is emerging as an IT-based core competence within the company.

9.5.4 IT users

Through our investigations, we found that in today's competitive and information intensive environment, construction managers are faced with more challenges when managing their organisation. This is partly due to changes in the organisational structures within companies, the nature of employment in the industry and the reality that IT users are often treated as second-class citizens. The increasingly decentralised models of management in construction companies require that IT users are educated, well trained and empowered to deal with IT issues at their level. They should be involved in the choice and design of IT systems, be allowed to participate in the process of decision making concerning the progress and management of IT and encouraged to seek innovative IT solutions in their specific areas. One commercial director interviewed commented, 'As a physical part of any IT system, the user is a critical success factor in improving the organisational effectiveness and much of the creativity that has come about through IT.' This reflects many of the points stressed in the chapter on implementing an IT strategy in Part A.

This user involvement was clearly demonstrated in three of the cases. We found that personnel from operating units within the business as well as administration and IT support departments were responsible for strategic IT initiatives from the bottom upwards. Industry suggests that there is growing understanding by construction organisations of the need for well-organised and managed training programmes. One company spent 15 man years of training when implementing their IT strategy. Although the extent and nature of training vary from one company to another, the benefits of such investments are clearly felt by all of the organisations we have studied. One group IT director suggested, 'We should pay 90% attention to organisation and 10% to the technical side of IT.'

IT improves the efficiency of the employees, and hence, the company

This view was shared by the managing director of a major building services contractor who said, 'IT systems are users' systems.' One CAD specialist within a major construction firm suggested:

> "Full integration of IT systems in construction companies depends on the integration of personnel with high IT skill levels and IT systems, therefore a substantial part of IT investment should be directed at personnel development programmes, if we are to benefit from the opportunities provided by IT."

Case study: Refurbishment modelling

A medium-sized US constructor has developed a highly innovative visualisation facility linked to project planning systems that enable the company to demonstrate to users the sequence and implications of refurbishment works on existing buildings, whilst still occupied and operational. This has been used to offer a differentiated service in new niche markets created by recent legislative change.

9.5.5 The strategy development process

Within the work in preparing this chapter, a number of examples of successful IT strategies were found. The best two IT strategies recorded were developed with a view to linking corporate, business and operational strategies together with a disciplined approach to strategic analysis and options. Contrary to the approaches adopted by some construction organisations, where the participation of consultants and end users is limited to the early stages of development, in these two companies the process of strategy development was viewed as a team activity. It benefited from the involvement of all the stakeholders, including suppliers at all stages. An elementary point about the development of IT strategies in these companies was starting strategy development from an initial suggestion of the best initial solution. This could then be subjected to rigorous feasibility analysis and at this stage a senior management champion from within the company is required. The companies then improved on initial proposals by taking the following steps into account:

- nominate people from various departments to participate in the process;
- analyse the company's current situation concerning IT;
- assess your human and IT resources and expertise;
- identify your key business activities;
- identify your key operational areas;
- identify key competitors and analyse their situation with regard to IT;
- discuss the outcome widely within the company;
- generate alternative strategies;
- consider options;
- formulate strategy;
- publicise internally; and
- implement strategy.

The above is a summary of the approach adopted by these companies. However, because companies differ in terms of organisation structure, management style, culture and values, we recommend that this process should be conducted with a methodological approach. This should be based on a clear budget and based on predefined expectations, realistic targets and a clear timetable with control and monitoring mechanisms.

Case study: Strategic IS planning

Changes in corporate culture in Japanese construction have been brought about by major government changes in industry practice triggered by international trade agreements. These have been exploited as an appropriate opportunity to make major new strategic investment and development in systems for client relationships and development as part of a strategy broadening out from a narrow focus on engineering integration.

9.5.6 The strategy implementation process

The process of strategy implementation is viewed by some construction organisations as the process of installing technology. In these companies, IT strategies are often dominated by technological issues. In contrast we recorded seven examples where the implementation process was based on a clear and well-defined organisational plan. Often, an IT professional was responsible for the overall aspects of the project whilst reporting to a non-IT senior construction professional and senior management champion. A modular mode of implementation often followed a successful pilot. When we analysed the evidence from these companies, we identified four interdependent stages:

- analysis of the existing IT facilities and resources, value added systems and their future requirements;
- planning and allocation of resources including an action plan;
- system installation and conversion and training programmes; and
- reviewing the overall process including budget, systems and attitudes.

The group operation director of one major consulting firm suggested, 'IT strategy implementation contains components such as new technology, procedures and rules that change the way construction companies manage IT and conduct business. Hence, the implementation of IT strategies should encompass the action from formulation of strategy and installation through to utilisation of the system and followed by evaluation.'

9.5.7 IT investment

There is a general view that IT investments often fail to deliver their expected benefits. We will return to this issue in detail in

Chapter 14. We found that the root cause of this is insufficient account being taken of the relationships between these technologies and the business and organisational context in which they are located. These fundamental problems are often experienced in the introduction and implementation of IT systems in all industries.

The critical factors for successful implementation of IT strategies are:
• the company's commitment;
• financial resources;
• IT resources;
• business benefits; and
• communications and participation.

We found that IT expenditure typically amounts to less than 1% of a construction company's annual turnover. This includes the cost of IT facilities, personnel, education and training, the latter accounting for a small proportion of the total investment. Profitable construction performers are spending a higher proportion of their annual turnover on IT and generally the companies who invest more on IT are more committed than their competitors, and, as a result, financial and market gains belong to them. These investments are often aimed at modern and up-to-date IT facilities, sophisticated IT systems, widely based IT training, skills development programmes, research and development, innovation and using IT to get closer to clients and stakeholders. There is also evidence that the absence of a proven economic model for evaluating IT projects and investments together with a lack of financial resources and long payback periods are often instrumental in the lack of progress of IT in construction organisations. A shift in thinking is required for senior managers in construction to better understand the strategic and value-adding opportunities that IT can give, rather than focusing too much on cost reduction and short-term returns on investments.

9.5.8 Who initiates IT investments?

IT investments should be justified by their value-adding potential

To meet with companies' corporate and business strategies, IT investments are often initiated in response to system inefficiencies, industry standards, client's requirements and competition. Major IT investments, leading to new and sophisticated IT applications, are usually initiated by IT, commercial, finance and managing directors. These investments take a corporate perspective compared to those investments in specific application areas, where marketing directors and other personnel from various operational parts of the business are often instrumental in initiating the investment. Regarding the latter, we documented six examples in the areas of marketing (three different areas of marketing in three separate case studies), design, industrial relations and health and safety. In these cases, due to their low-cost development, the initiatives required very little in terms of justification for investment but resulted in important strategic and value-adding benefits.

Case study: Cash flow management

The Cash Flow Management System (CFMS) is a dedicated system, developed by the company for the efficient management and utilisation of project finances. It is also designed to monitor the cash flow and to report on projects at divisional level and to provide the group management with financial performance related information.

9.6 Summary

Whilst a small number of construction organisations have managed to achieve business efficiency, competitive gains, market leadership and a healthy return on their investments, on the whole, the management of IT in construction organisations is in need of more attention and there is considerable room for improvement. The industry suggests that IT investments are often accompanied by poor vision and implementation approaches, poor planning and coordination and little adoption of long-term IT strategies linked to and supporting business strategies. The reasons for this include a series of organisational and technical problems that vary between individual organisations. These problems constitute barriers to construction companies seeking to benefit from the potential opportunities offered by IT. They result in more time and effort being required for the effective and successful management of change.

There is also clear evidence that strategic advantages and opportunities provided by IT have not always been sustainable in the long term, requiring continuous strategic re-evaluation if advantages are to be long lasting. Some examples of strategic IT exploitation arise as a consequence of specific R&D and innovative approaches to IT investment and development. However, the exceptional examples are few and we would recommend that a more widespread link be formed between companies and research organisations.

Case study: Contract control

A fully integrated contract control system was designed and developed by a leading building services company to monitor and control each of the individual components of a project together with practical applications of control techniques of progress, variation, quality and coordination.

9.6.1 Key propositions

Although it may be true that the progress in the use of IT in construction organisations is slower than that in other sectors of the economy, industry suggests that, despite the influence of adverse internal and external factors, some construction organisations have made significant progress. However, for other construction organisations to follow their example, industry has found the following key ways of moving ahead.

- Companies must harness IT in the direction in which their mission is driving the organisation into the future.
- The involvement and participation of IT professionals and senior construction professionals as IT directors in the formulation of corporate and business strategies is necessary.
- The creation of an IT infrastructure is fundamental.
- The development of a new IT culture based on well-defined and organised education and training programmes for all IT users is critical.
- Encouraging and fostering participation and empowerment at all levels in IT related matters encourages new initiatives to emerge and encourages support for strategy development and implementation to be gained throughout the organisation.
- Develop a mechanism for the justification of IT investments based on value-adding and strategic criteria.
- Recognise that IT is a resource that needs to be planned, monitored and controlled.
- Companies must encourage innovative IT approaches for efficiency, client satisfaction and problem solving.
- Give careful consideration to operational and functional integration, involving the integration of business, users and IT domains.
- Gain an awareness of the latest technological developments through more active participation with research and development and innovation.

9.6.2 How to be strategic

The industry is saying that to be strategic in the use of IT means to develop a business strategy supported by IT that would deliver better services and products than competitors. The main issue concerning IT strategy is that it is extremely difficult to forecast future developments in this area, yet is not good business practice to stay behind. Therefore the recommended strategy may be the one which is successfully adopted by some companies, namely that construction organisations should scan the environment for the latest current developments that can be adopted for their specific business, environment and competitive opportunities. This philosophy also suggests that the combinations of short- and long-term approaches often lead to a successful outcome with cost advantages, market gains and a positive impact on corporate managers and end users. The important issues we recorded concerning 'how to be strategic' are summarised and presented as the following questions that all companies should ask of themselves.

- Where do we make our money?
- Where are we in relation to our competitors?
- What do our clients think?
- Where are we going as a business?
- How can IT add value to our organisation and business activities?
- How can we achieve that added value?
- How do we implement that and how much should we spend?
- How do we control and evaluate IT initiatives?

10 Benchmarking as a methodology

Brian Atkin, Martin Betts and Andy Clark

"IT improves the response time and business efficiency."
– *UK finance director*

A major technique of strategic management that has gained in popularity recently is benchmarking. There is a strong case for the more widespread use of benchmarking in construction. There are various types of benchmarking that can serve various purposes. Benchmarking can focus on a range of business process issues, one of which is IT.

The origins of benchmarking lie in the quality movement. Benchmarking of processes between industry sectors has been successfully applied in banking, airlines, the hotel sector, retailing, insurance, engineering and many others.

- Types of benchmarking include: internal comparisons between parts of the business, competitive comparisons within a sector and best-practice benchmarks against best-in-class companies.
- The benchmarking process embraces: identification of the core issues, internal and external data collection, analysis and implementation of change.
- The management issues on which benchmarking can focus include: visions, plans, justifications, implementations, controls, organisations and learning.

"The problem isn't just data processing, but one of managing a project which changes daily."
– *Operations director, Belgian subcontractor*

10.1 Introduction

This chapter focuses on how IT can be benchmarked to provide construction companies with essential information of how their competitors and out-of-industry organisations are using technology to support profitability. Benchmarking as a concept identifies with the practicality that business organisations do not have the luxury of time when organisational improvements are required in order to remain competitive. Marketplaces are currently subject to the competitive forces of flexibility, speed, quality and cost. These forces are not just influencing the construction market, but every market, and real lessons can be learnt from how other industries have met similar demands. We have argued in earlier chapters that construction companies cannot ignore the rapid pace of market, technological and organisational change through examples of strategic planning methodologies. This chapter argues the need for construction organisations to benchmark against how world-class organisations use IT to support profitability and competitive advantage. The case for action looks at examples from a variety of industries with the salient point being that in order to ensure success, the benchmarking initiative must be business- rather than technology-focused.

The need to benchmark IT stems from the link between the rapid change in the technological infrastructure of the world and the changing methods and assumptions of how business is conducted. Information technologies have enabled communication across international boundaries to be instantaneous and storage facilities of enormous quantities of information allow ready access at a steadily reducing cost. Powerful database applications have facilitated the essential capability of sorting data to overcome an increasing information malaise as business managers are faced with a deluge of paper and e-mail to read. It is clear that electronic communication infrastructures are a prominent part of the way some companies conduct their business and the only realistic future for the remainder. Alongside advances in IT are advances in philosophies of business governance that break the managerial norms of the previous century and harness current emerging technologies. Even the most cursory glance through the management press will reveal myriad articles discussing many of the concepts introduced in earlier capters. Technology and management should not be viewed independently as the synergy between the two is their strength. A fundamental justification for BPR has been the growing disillusion among the business world with the promises from IT pundits who have been enthusiastically forecasting that the IT revolution is upon us with a correlating effect on business performance. It is clear that deploying technology to a business problem will not penetrate organisational performance unless the business structure and process is redesigned accordingly. Business managers have deployed the latest technology, hoping to solve the faults of a poor process. In reality the same bottlenecks still exist, we just reach them quicker. Unfortunately BPR is a high risk activity because the clean sheet approach to process redesign necessitates a blank cheque to implement, with a high level of uncertainty regarding the outcome. It is difficult to make a business decision on this basis. Success stories such as those of Ford Accounts Payable (Chapter 4), Mutual Benefit Life and IBM Credit (Davenport, 1993) seem to have become the exception rather than the rule, resulting in the need for a less haphazard approach to organisation rejuvenation.

10.2 The case for benchmarking in construction

Existing in such a dynamic environment means that forecasting and managing change become key skills for business managers. Responsiveness to customer needs and desires allows the world-class company to stand out from the crowd of competitors. To stand out from the crowd forces a different way to play the game, a new basis to conduct business. Managers in successful businesses are becoming increasingly armed with a variety of tools that allow their businesses to be better than the rest. De Gues (1997) studied the lifespan of companies in the western hemisphere to find that the average corporate life-expectancy is below 20 years, while only the larger companies survive between 40 to 50 years. The study found that the average life-expectancy of a company is well below the potential life-span, as there were a number of companies whose existence had spanned centuries. In focusing on a group of 27 companies who were older than a century, their main strength was the ability to manage change. There were four further correlations which were conservatism in financing, sensitivity to the world around them, awareness of their identity and tolerance to new ideas. It would seem that the odds are against any company expecting to survive beyond 50 years without taking action to plan and secure that survival.

Planning for survival must involve a cognisance of the operating environment. As we have seen elsewhere in this book there are many ways to assess a market and even more to lose market share to a competitor. The concept of benchmarking was developed from the experiences of a company facing a dramatic loss of market share. The Xerox Corporation faced a competitive threat in the late 1970s from Canon who could make, ship and sell comparable photocopiers for less than Xerox's manufacturing costs. Xerox, like many Western businesses, failed to perceive the potential threat of Japanese industry until that threat imposed a competitive challenge that could destroy the business. Japanese industry had spent many years building itself into a global force based on the continual improvement techniques enshrined in *kaizen*, the gradual, unending quest for improvement by doing things a little better and setting, then achieving, higher and higher standards. Instead of relying on internal task forces to identify process improvements, Xerox looked externally to companies who were regarded as world-class in their performance of a process that they had identified as mission-critical. By studying how and why the external companies performed to a world-class standard, Xerox were able to learn and replicate operations that allowed them to cut costs by 50% and lower defect rates to Japanese standards.

As a process improvement tool, benchmarking recognises the problems companies such as Xerox have in maintaining a competitive edge in a dynamic marketplace. A reliance upon the replication of internal best practice forces companies into myopic management practices that fail to recognise a panoptic competitive environment. The response developed by Xerox has been to study the internal and external practices of efficient companies with a view to learning and implementing superior processes that will contribute to a strategic competitive advantage (Leibfried and McNair, 1994). Many companies have been unable to change themselves sufficiently quickly against competition and have subsequently become mere legends of bygone industry. IT has a very specific contribution to show what changes are possible. By examining how other, well-respected industries have employed IT in some detail, opportunities will be presented to the construction industry. Benchmarking is a response that has a well-developed management toolkit for identifying the processes that will migrate between companies and for providing a methodology for enacting the implementation of the improved process.

In all cases a benchmark must be focused on setting objectives and goals which ensure competitive excellence strikes at the heart of performance measurement. Benchmarking is the means for identifying performance levels and provides the basis for continuous improvement. For it to work successfully, benchmarking has to be stakeholder driven, forward looking, participative and focused on quality. A consequence is that it forces issues out into the open so that they can be dealt with according to the goals of the company. It identifies current practice and through external comparison identifies best practice and the actions that are needed to match and exceed that best practice. By bringing in the external dimension, energy can be channelled away from internal conflict towards a focus on achieving the competitive edge based on providing an excellent service to customers.

During the 1990s, businesses have increasingly focused on the needs of their clients and customers and the extent to which they satisfy them. Whilst this is clearly necessary, it is not everything. Clients exercise primary interest in the end product or service and companies look upon their suppliers as the means to help deliver that end product. The point is, however, that businesses need to consider their core processes if they are to deliver against increasingly demanding client and customer needs (Hammer and Champy, 1993). In this context, it is important to consider all organisations and people who have a stake in the survival of the company. These stakeholders, as they are known, include owners, employees, suppliers, and customers. They depend on one another, and because relationships consume resources, costs are incurred. Management of these relationships is necessary from an accounting perspective since cash flow is the lifeblood of most organisations. Benchmarking can provide management with the tools needed to make decisions about policies and practices with regard to the most fundamental processes of a construction company. Managing the life cycle of a building has to become more competitive if companies are to thrive. Clients are looking for greater efficiencies and lower costs and the construction industry will have to respond if it is to survive. Future survival will depend upon the integration of the construction process as the interfaces between stages in the construction process and between companies are a primary cause for inefficiency. As soon as information is subject to a hand-off between parties, much of the knowledge that went into that information is lost. The construction process has a preponderance of information hand-offs between different professions, usually between different legal entities. With an adversarial culture this reluctance about team-working culminates in extensive recourse through litigation when disagreements cannot be resolved. The legal framework that governs the construction process promotes the interest of the individual company, but not the multi-party project, which is the interest of the client. Petrozzo and Stepper (1994) see hand-offs of part-completed work as a major failing of most processes and advocate a team-centred approach that follows the value-adding process structure of an organisation. The implications for the construction process are that all the parties to a project work as a single team to ensure that information and knowledge are captured and communication is facilitated. Whether this works from a common location, or by enabling technologies, there are other industries who have faced and overcome identical problems with proven solutions that work.

Unfortunately the construction industry claims a very special and distinct way of operating which does not lend itself to learning from other industries or indeed from within itself. The pivotal role of the client in the design of the end product and the well established processes within the industry prevent the degree of flexibility necessary for the industry to change radically. If this is true then existing players in the industry will be under pressure as competitors enter the market offering a service that customers demand, rather than being forced to engage the services of an industry generally regarded as a necessary evil. Clients only proceed to design and construction of a new facility if they have no other option. To accommodate a function within existing facilities is a quest that is spared no effort. Alongside the desire to not use the construction industry is the rationalisation of many blue-chip businesses that has reduced built assets and increased the use of IT for hot-desking and home-working, all of which have reduced the amount of space an organisation needs to occupy to go about its business. While this has affected private and public office space, similar changes are underway to our infrastructure requirements with the laying of information super-highways instead of physical highways.

Benchmarking inside the construction company, the industry and outside the industry may prove to be fundamental to the survival of many companies who operate in a sector better known as the last of the large industries to face the change proposed by IT. In continuing to procrastinate by ignoring IT as a competitive weapon, companies within the sector must face the reality that they are highly susceptible to competition from all directions. We can no longer ignore the chameleon-like ability of companies to apply core skills to different markets. An example is Virgin who have moved from a record label to expand into an airline, a hotel chain, a soft drinks producer and, recently, a financial services and railway operator by actively marketing an image that appeals to a demographic group. If the time is looming when companies, attracted to the size of the construction market, apply superior logistical, fabrication, organisational and project management skills to the construction process, then action needs to be taken. And if this is not the case then internal competition should be enough to stimulate action. IT with the appropriate managerial philosophies can provide the answer if deployed correctly. The case for benchmarking is that there are a large number of companies in a variety of industries who have already invented that wheel; construction companies have the choice of learning from others or embarking on the painful path of self-discovery.

10.3 Types of benchmarking

Benchmarking is traditionally a one-on-one activity, although the following chapter examines the results of a construction-specific benchmark that benchmarked a group of companies against world-class comparisons. There are several types of benchmarking and Lewis and Naim (1995) identify four types: internal, competitive, parallel industry and best practice.

Internal benchmarking

Internal benchmarking is often the first step for a benchmarking project as an effective method of climbing a learning curve. It is often very revealing. Internal benchmarks subject the firm to the process of benchmarking across divisions or departments that perform a similar or identical process. By nature they are easier to complete than external benchmarks and are often a source of surprise due to variations across different parts of the company. In particular, construction companies have much to benefit from an internal benchmarking exercise due to the sometimes large number of projects that often have wide discrepancies in performance and innovation. An internal benchmark solely focuses on the source company of the benchmarking initiative and provides the basis for the comparison with external companies.

Competitive benchmarking

Competitive benchmarking occurs between firms within the same industry sector who sell an identical or similar good or service. This form of benchmarking is often the most difficult as it relies upon competing firms to share details of how processes are performed with each other and the approach used should focus on operational aspects of process performance rather than strategy. Chapter 11 describes the results of a competitive benchmark between different groups of companies within the UK construction sector.

A competitive benchmark is best administered by an independent party for success. This allows the exercise to overcome the potential suspicion and mistrust that can occur between competing organisations. An anomaly is the United States Air Force who realised that their internal training processes were inferior to those of other armed forces units. By developing a benchmarking matrix of 13 or 14 important areas for success, they were able to assess themselves against other armed forces units in the USA and UK. Lessons were learnt that destroyed many of the basic assumptions they were using as operational guidelines to train new recruits. As a result, new recruits had shaved a minute off a two-mile run and increased participation in the classroom by 312%. Many other operational successes were seen that made US Air Force standards comparable to world standard.

Parallel industry benchmarking

Parallel industry benchmarking occurs between companies from different sectors who undertake a similar process of production or service. This type of benchmarking is considered easier than the previous example as issues of access and willingness to participate in a comparative study will not be as problematic between companies who are not in direct competition.

Best-practice benchmarking

Best-practice benchmarking considers the merits of a comparison from a particular market leader who is known to have an exemplary process that is similar to the process under study. While all of the operations of the process may not be totally transferable between firms due to different industry structures, there will often be important lessons that can be learnt.

Best-practice benchmarking is called innovation benchmarking by Davenport (1993) as the target processes are often born out of innovative thinking and bold managerial implementation. Hammer and Stanton (1995) conversely argue that benchmarking stifles innovation and should be used with caution, an argument that is acceptable if you have a blank cheque for the implementation (Davenport and Stoddard, 1994). Very few companies can afford to be genuinely innovative when approaching process redesign as unknown costs and benefits do not make a good case for business planning. The innovative process may be easily implementable and provide the rewards that theory promises, but it may be dominated by problems with associated costs that outweigh operational benefits. Instead, choosing the path of a quantifiable and proven process provides a template for improvement, based on real experience and measurable progress. Camp (quoted in Liebfried and McNair, 1994) argues that the innovative nature of benchmarking is in the implementation of the best-practice process where there is usually opportunity to adjust the operation of the best-practice process to better the operation of the source company. Innovations have originated in the migration of processes between industries that change the basis of competition within the industry. Dell Computer changed the sales and distribution standards of the personal computer industry when it offered direct mail order to its customers as a method to purchase a computer. Mail order is a well established sales and distribution method, yet when applied to a different industry using benchmarking as a tool to learn from others, the result is an innovation. Finch *et al.* (1996) discuss how bar-coding can be used within the construction industry by using the retail industry as a blueprint for development. Application to date has been limited, but the possibilities are considerable as an on-site data-capture and logistics tool.

10.4 Benchmarking methods

Before embarking upon benchmarking methods some caution needs to be considered. Successful strategic management relies upon component processes being sufficiently coherent to support the aims of the company (Davenport and Short, 1990). Processes have to be designed with care to ensure that the chosen combination supports the strategic vision: contradictions should be rectified before they are operational. Without caution, benchmarking could lead to a mix and match approach to improvement with the potential consequence of being strategically 'stuck in the middle' (Porter, 1985). Care should also be exercised to choose an analogous process to benchmark. Leibfried and McNair (1994) advocate process mapping and the development of an activity grid to define the salient features and the economic or regulatory constraints of the process for benchmarking.

Benchmarking, in the strictest sense, is a one-on-one activity where two companies perform an open-book exercise about how they each perform a target process. Despite this, benchmarking is equally applicable to groups of companies, as shown in Chapter 11. The former highlights the detail of how a company may perform a process, while group benchmarking can highlight both detail and industry trends. To a company taking part in a benchmarking project the approach is similar whether one-on-one or as a member of a group, but group benchmarking is more effective when facilitated by an independent third party such as the work of the IMVP (International Motor Vehicle Programme) (Womack *et al.*, 1990) or the Construct IT Centre of Excellence, as explained in Chapter 11. A high level diagram of a generic benchmarking process is shown in Figure 10.1.

Stage 1: Identify core issues

A company's structure may include any number of processes although a sensible breakdown of processes for focus is advisable. Davenport (1993) advises that between ten and 20 processes will enable a company to understand each process, with each being of a manageable size to redesign. Processes, from the discussion of BPR in Part A where we stated that they should be value-adding to internal or external customers, are the basis by which a company creates wealth. The salient point of identifying core issues is that they should have a high impact on the performance of the company, or be mission-critical to future business. Benchmarking a process with little strategic or operational impact will have little effect on bottom-line performance. Process mapping is a useful exercise at this point as it should allow an understanding of the current process structure of the organisation. Strategic planning techniques, as explained in Chapter 2, are commonly used to focus on core issues.

Stages 2 and 3: Internal and external data collection

The quality movement has had a significant influence in the use of performance metrics to support management decision-making. This approach has been correspondingly important to benchmarking methodologies in providing a baseline to benchmark against an internal or external partner. Most of the well pub-

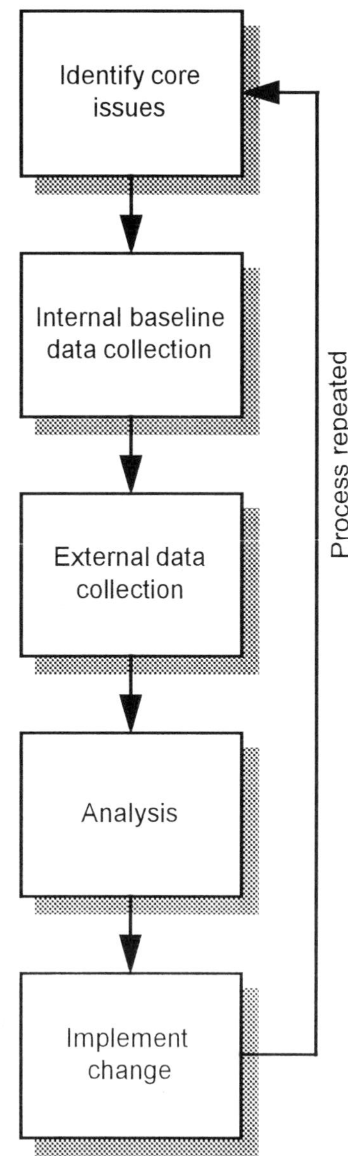

Figure 10.1 A generic high-level benchmarking process (developed from Liebfried and McNair, 1994)

licised management success stories have been littered with metrics that illustrate process improvements. Davenport (1993) notably uses:

- IBM Credit, who reduced the time to prepare a quote from seven days to one day;
- Federal Mogul, who reduced the time to develop a new prototype from 20 weeks to 20 days;

- Mutual Benefit Life, who halved policy underwriting and issuance costs; and
- the US Internal Revenue Service, who now collect 33% more revenue from delinquent taxpayers, with half the staff, resident in a third fewer branches.

While it is important to base benchmarking data collection on a metric-based method, it is inadvisable to found the data collection process on metrics. Without an understanding of the process under scrutiny it is difficult to make a serious judgement of what the metrics actually mean. Benchmarking IT is very susceptible to bean-counting, making judgements based on level of investment, number of PCs per user, the particular software being used, etc. But to effectively benchmark, a balanced understanding between how fast or cost-effective a system is and an understanding of why the system is fast or cost-effective is essential. The blend of quantitative against qualitative data is an important consideration in designing the data collection methodology. The book *The Machine That Changed the World* (Womack *et al.*, 1990) describes, with a mix of metrics and detailed commentary, an in-depth study that showed how Japanese automotive manufacturers could build cars quicker and with a lower defect rate than North American manufacturers.

Research methods vary from questionnaire, survey, case study, action research, etc. Without entering a detailed discussion of research methods which is well provided elsewhere, we consider that the following guidelines are important to collecting the highest quality data.

- **Anonymity.** Interviewees should be granted anonymity from their input to the research process. Attributing people to comments will hinder honesty, forcing interviewees to reflect what they think managers would want them to say, rather than what they think.
- **Cross-functional knowledge.** Interviewing a variety of job profiles is important to understanding a process as employees with differing functional skills see the process in a different way. One of the foci of human resource structures in a redesigned process environment is that multi-functional process teams mirror the operation of the process. Production and marketing professionals work with design professionals to assess if a design can be built and sold from the earliest stages, alleviating risk proactively. Upstream skills have an important bearing on

the downstream process and a benchmark should recognise the value that those indirectly involved in a process can have.

- **Interview the best people.** Benchmarking suffers from the situation that to be effective the most competent staff should be interviewed, exactly the people who are required for every other important task within the organisation. The danger exists within a benchmarking initiative that due to time constraints the most able staff will be unable to feed into the process, losing important knowledge and insight. As a result, management support must be used to ensure that key people are given the priority and time to participate in the effort.

Stage 4: Analysis

Analysis of the results of a benchmarking study should look to develop causal relationships between performance and operation. Metrics will have been collected to compare the operations of two or more processes in terms of cost, quality and productivity. Assessing the significance and meaning of these variations is the task at this stage of the benchmarking process.

Researchers commonly mistake causal relationships, highlighted by Elton Mayo's Hawthorne experiments (Pugh, 1971). A team of researchers sought to relate working conditions to productivity at the Western Electronic Company's Hawthorne plant. Results showed a steadily increasing rate of productivity when working conditions were changed for the better or worse. When the research focus changed from a Tayloristic scientific management approach to a human relations approach the team found that the increases in productivity were mostly due to team dynamics and partly due to the interest shown by the research team in the workers. Data analysis has to be approached with objectivity. It is a simple task to skew statistics to prove a theory.

The analysis of the benchmarking data also provides a plan of action for the implementation of the best-practice process. The DTI (1992) advocate the use of a benchmarking matrix to display the data analysis. A table showing actual performance against target performance, correlated against the performance of competitors, places a firm focus on the degree of performance improvement required.

Stage 5: Implement change

Change management is probably one of the biggest challenges facing an organisation when benchmarking. All of the work prior to implementing the benchmark will stand for nothing if not implemented, and research suggests that the majority of change management initiatives fail at implementation. Failure statistics for BPR projects have been as high as 85%, which has dampened the enthusiasm generated from the success stories of Ford, IBM and Hallmark (The Economist, 1994). Analogies can easily be drawn to benchmarking projects, as an equivalent magnitude of change is imposed on an often unwilling workforce. Benchmarking, however, differs from a BPR exercise in that an operational model of the process exists in the benchmarked company. A further benchmark that could assist the change management process is to examine the pitfalls experienced by the benchmarked company when they redesigned their processes.

Change management is concerned with preparing an organisation for closing the gap between current processes and the best-practice model. It is a part of organisational operation that needs careful attention, as it can be managed. Turning a seemingly unwilling workforce into a willing workforce is a difficult task, but the following points are considered essential to a change management initiative.

- Leadership from the most senior levels of management is essential. Having the seniority to force through change when required, alongside a passion and belief in the necessity of change, will overcome problems that may seem insurmountable to those without the clout in the organisation.
- Project management skills that ensure that the project remains on track throughout to a rigid and structured plan.
- Communication to all employees involved from the initial stages of the project, focusing on imparting accurate information. Stopping rumours before they start can only be achieved with jargon- and cliché-free communication through a variety of media.
- Buy-in from all employees involves more than a superficial nod of agreement when faced with a senior manager explaining the need to change. Buy-in ensures that employees feel part of the process and can be achieved by using their input, listening to ideas, grievances, comments and giving responsibility for parts of the process. The North American Coal Corporation conducted a pilot study of a focus process to show the potential for improvement. The pilot study cost $80 000 and saved $500 000. Proof of success is a powerful tool to persuade sceptics.

10.5 Benchmarking focus

Benchmarking as an organisational improvement tool can be particularly powerful if used effectively. As we have discussed, the impetus for change is currently driven by the opportunities afforded by IT, a change that, if managed correctly, can bring dramatic increases in business performance. Benchmarking has been shown to be a management tool that shortcuts the change management process by providing a framework to learn from organisations who perform to a standard that is regarded as world class. A benchmark of IT is therefore an important issue to a construction organisation and, unless managed properly, will not provide the meaningful results to change the operation of the organisation. Above all, a benchmark of IT use should focus on business issues and not technology.

Much of the early focus on IT as a competitive weapon has been superseded by the BPR viewpoint of imposing massive change on an organisation as a cure for the functional structures that have dominated as a legacy from the work of Frederick Taylor and Adam Smith. Proponents of the strategic management of IT school argued that organisations should take a business advantage from emerging forms of technology, exploring ways the technology could fit into the business. Earl (1989) has questioned this philosophy, observing that business organisations have initiated many projects that were based on implementing a technology to bring about a competitive advantage, in most cases being 'oversold and underdelivered'. Many writers, notably Scott-Morton (1991), have discussed the all-pervading ability of IT to affect every part of an organisation's operations. The underlying problem seems to be that the focus of a business improvement project as IT-led has divorced the technology from the business issues it is intended to address. The result within the organisation is that the technology is managed by the technologists and misses the fundamental needs of the business or is not supported by the rest of the organisation. The IT director has everything to lose while business managers sit back and allow the project to fail. Earl (1989) observes that an IT project is a project doomed to failure, while projects that were initiated to address a business problem, that culminated with IT as the key enabler, were successful. His framework for a change of focus is shown in Table 10.1.

The framework tracks the change in philosophy as we have progressed our understanding of how IT can be exploited by business organisations. Imposing this philosophy on a bench-

Table 10.1 Putting business back into IT (Earl, 1989)

Management issue	From IT	→	Business change
Vision	Technology futures	→	Rethinking business
Planning	IT strategies	→	Business themes
Justifying	Financial appraisal	→	Business case
Implementing	Project management	→	Managing benefits
Controlling	IT expenditure	→	Cost of business
Organising	IT business	→	Business IT
Learning	IT literacy	→	Organisational development

marking analysis is an essential step to benchmarking business ends, instead of IT means. The outward motive for benchmarking has to be improvement for the organisation; anything else, as a number of observers have noted, will result in sub-optimisation of a methodology that, if applied correctly, has the power to afford dramatic business improvement. Osterman (1991) found in a benchmark study of North American automotive manufacturing, that plants subject to high technological investment, without the corresponding change in human resource policies, experienced no significant change in productivity or defect rates. Conversely, plants with moderate technological expenditure and sophisticated human resource policies significantly improved productivity and defect rates. Business-led change has to touch all parts of the organisation to be effective; attempting to isolate certain areas will not affect overall performance.

10.6 Summary

The construction industry must accept that it has to change. It has been long characterised by offering a service that rarely meets the clients' needs of cost, quality and time. As it now faces an increasingly demanding client base who require the industry to add value and not cost, the inability to accept this will result in a similar fate as the UK automotive manufacturing industry, either forced into bankruptcy or taken over by overseas corporations who see the potential of the sector and who realise the impact that efficient management will have. Evidence of both is already visible in UK construction and if left unchecked, the trend looks certain to continue. The UK construction sector, therefore, has two choices. It can either accept its fate by continuing to perform in a mediocre way, or it can fundamentally change its operations and become a sector dominated by world-class process management.

The opportunities afforded by IT and process-led change offer the UK construction industry the ability to become a world-class sector. Concentrating on value-adding business processes enabled by IT has proven key to the competitive survival of a number of other industry groups, most of whom have suffered the hardship of dramatic organisational change. The construction sector can either benchmark against those other industries or embark on its own path of discovery, experiencing change the hard way. To take the latter option usually involves the implicit assumption that the operations of the construction industry are so far removed from any other sector that benchmarking against them is a pointless activity. While a building cannot be constructed the same way that a widget is manufactured, there are lessons that are transferable between the two worlds that benchmarking sets out to exploit.

To use benchmarking as a change management tool for the implementation of IT demands the necessary focus. The objective must be business improvement rather than technological deployment for any chance of success to be guaranteed.

11 Benchmarking IT use in construction

Brian Atkin, Andy Clark, John Gravett, David Smith and Stephen Walker

"We are only just starting to use the tip of the iceberg compared with how other sectors benefit from advanced IT."
– Managing director, UK special projects contractor

Benchmarking of the use of IT to support discrete construction processes has been applied to more than 70 UK construction businesses. Comparisons have been drawn with other industries including automotive, manufacturing and engineering. Qualitative comparisons of typical construction practice and world-class best practice show the extent of the gap to be bridged in industry improvement. These gaps are at their greatest in site processes and managing supply chains. The gaps are narrowest in cost estimating and bidding, and cost and change management. These results have implications for individual firms, the whole industry, and for research development.

- IT benchmarks have been completed in: briefing, design, cost estimating and bidding, cost and change management, construction site processes, supplier management and facilities management.
- Out-of-sector comparisons are drawn with: automotive component supply, automotive manufacture, nuclear decommissioning, real estate management, engineering production and general manufacturing.
- Use of IT in construction lags best practice in other sectors.
- Variability in IT use among construction companies is very high.

"Where systems aren't compatible they tend to go back to the good old hard copy for information."
– Project manager, UK contractor

11.1 Introduction

Benchmarking, as we have seen in the previous chapter, has attracted much attention over the past few years as a tool of business process redesign (BPR). That said, there are a few examples of successful benchmarking in construction. This chapter will report on the findings of one of these examples - a series of benchmarking studies on the use of IT. The studies form part of a UK national initiative aimed at improving the application of IT to construction, sponsored jointly by construction industry firms and the government. Briefing and design, construction site project management, supplier management, cost estimating and bidding, cost and change management, and facilities management have provided the subjects of study for which more than 70 of the UK's leading firms took part. In addition, out-of-sector comparisons were made against world-class best practice in each subject area. This aspect, in particular, provides valuable insights into how construction firms can achieve higher levels of performance. They must, however, clarify their business objectives, adopt radical policies and commit themselves to implementation of IT on a major scale. Those that do so will be amply repaid, in much the same way as those out-of-sector exemplars of best practice used in the comparisons. Details of the research and development, which must now follow, are also presented.

Benchmarking is a much talked of technique, yet one where relatively few results have emerged to benefit industry. As a tool of BPR, it can be used to ascertain current performance levels. When extended to a comparison with best practice, it can quickly indicate where and how performance can be improved. That said, few examples of successful benchmarking are evident in construction; fewer still exist in relation to the use of IT in these domains.

11.2 The Construct IT Benchmarks

For the purpose of the studies, IT was deemed to cover all electronic systems for capturing, storing, manipulating, retrieving and transmitting data and was not confined to computer-based equipment. The studies were based, for the main part, on structured interviews with senior managers, responsible for the domain of study and for IT, in more than 70 firms representing a mix of construction companies, designers and building owners. A questionnaire, covering the relevant stages or primary tasks within each mission-critical business process and containing typically in the region of 2000 individual questions, was drafted by the authors and tested on an industry-based steering group. Each questionnaire was then piloted with three firms and the feedback used to amend the form and content of a few questions. All remaining firms were then asked to participate in the main studies by selecting their best project or operation, within each process area, from an IT perspective.

The questionnaires were based on a generic model of the domain to be studied. In most cases, several clusters or related applications within the overall business process were identified. Questions were divided into those associated with the use of IT for a given activity and those signifying the importance of the application of IT to that activity. Scoring questions were included in all sets of activities and processes. Typically, these took the form of degree of IT use and a rating of importance of IT. Open questions accompanied each set so that qualitative comments and other factors could be recorded. Apart from answers to predetermined questions, interviewees were encouraged to offer comments wherever they felt these would add something to the quality of their answers.

All studies included a core of questions related to organisational aspects. These were grouped under strategy, policy and procedures. They considered, *inter alia*, the business and operational benefits of IT, basis of the decision to use IT in the designated process, extent of intra- and inter-organisational electronic flows of information, permanent or special communication links, networking of IT, benefits delivered to clients, responsibility for IT, access of staff to IT and reasons for any non-use.

The approach has relied upon self-evaluation by the firms in responding to the questions put at interview. Under this arrangement, it is always possible for a firm to offer less than honest answers. The authors are confident that the answers are a true and accurate statement of the firms' use of IT and belief in its importance in the context of the relevant business process. Since the purpose of benchmarking is intended to lead an organisation to identify and adopt improved business processes, distorting that assessment would not help in the search for improvement.

Whilst the results are useful for the firms, and others with an interest in the domain being studied, they are indicative only of current practice within the group of participating firms. A more meaningful approach and one that is implicit in benchmarking is to compare performance with another organisation, outside the domain of study, that is regarded as achieving best practice. By doing this, it is possible to measure the distance by which the participating firms fall short of best practice and to provide insights which might help them close that gap.

11.3 Out-of-sector comparisons

Selecting an acceptable example of best practice for comparison is one of the most difficult aspects of benchmarking. Often, people will hold strong views about what constitutes a valid comparison. Whilst similar organisations can be compared and measured against a notion of what constitutes best practice, this cannot take the place of real examples. Accordingly, the authors and the respective industry steering groups spent a good deal of time addressing this issue of comparisons, always mindful of the concern that any comparison would have to be defensible. That said, if the same or similar processes exist in an entirely different industry and if there is something that can be learned from best practice, there is no reason to avoid the comparison. The decision was taken to consider the merits of a number of diverse industries as objectively as possible.

Table 11.1 Choice of out-of-sector comparison for each study

Study	Out-of-sector comparison
Supplier management	Automotive components
Construction site processes	Nuclear decommissioning
Briefing and design	Automobile manufacture
Facilities management	Real estate management
Cost estimating and bidding	Engineering production
Cost and change management	Manufacturing generally

11.4 Primary findings

Responses from the interviews, as recorded in the questionnaires, were analysed, and a simple scoring system applied to questions covering the key issues. Scores were produced for each set of activities relative to use and importance. Additionally, the extent to which activities and processes are integrated have been determined. Scores have not been weighted.

The research team had previously considered carefully what might constitute best-practice IT. After much discussion, it was decided that best practice should equate to the electronic exchange of information and integration of data in support of the business process being studied. These factors were reflected in the maximum score attainable under the appropriate questions. The incorporation of world-class, best-practice IT in the studies provides a robust benchmark for overall comparison and was confirmed as achieving maximum scores in each of the studies.

Results for each set of activities in each benchmark have been plotted on a scatter graph, an example of which is shown in Figure 11.1. Maximum possible scores indicate best-practice IT, as provided by the out-of-sector comparison. The same maxima have been used to fix the relative positioning of all firms. Best-practice IT is identified by the maximum points attainable on both axes. Additionally, the area covered by the axes has been separated into four equal quadrants to enable system development and research priorities to be assessed.

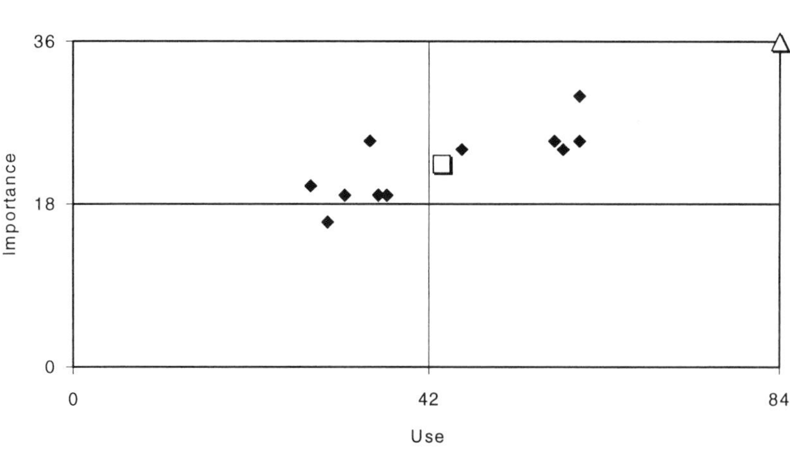

Figure 11.1 Performance graph showing the relative positions of benchmarked firms and the industry average, □

Firms are not identified by name, but each has been able to establish its position relative to others by inspection of the resultant graphs and accompanying statements and comments. Additionally, each firm has been given, in confidence, a marked up copy of the results showing its position.

The summary tables for each of six discrete process benchmarks, carried out to date under the Construct IT Centre's work, are reproduced in Figures 11.2–11.7. These are each followed by an overall interpretation of the results and then by Tables 11.2–11.7, comparing descriptions of the world-class best-practice company and of typical mid-range practice in construction.

11.4.1 Supplier management

Interpretation of overall performance

The distribution of scores in Figure 11.2 shows clearly that the use of IT in support of supplier management varies widely among the companies. For a group of four companies, the gap between their current level of use of IT and best practice is such that they might reasonably expect to close that gap in time. That said, the gap between them and one or two other companies is greater. This suggests that those companies with a lower penetration of IT have much to do if they are to move in the direction of best practice. The variability in rating of importance is small and suggests that the companies generally regard IT use in supplier management as important. For many it is the case that they use IT to the extent that they see it as important.

Qualitative descriptive comparisons

A comparison of best-in-class supplier management with mid-range construction performance is given in Table 11.2.

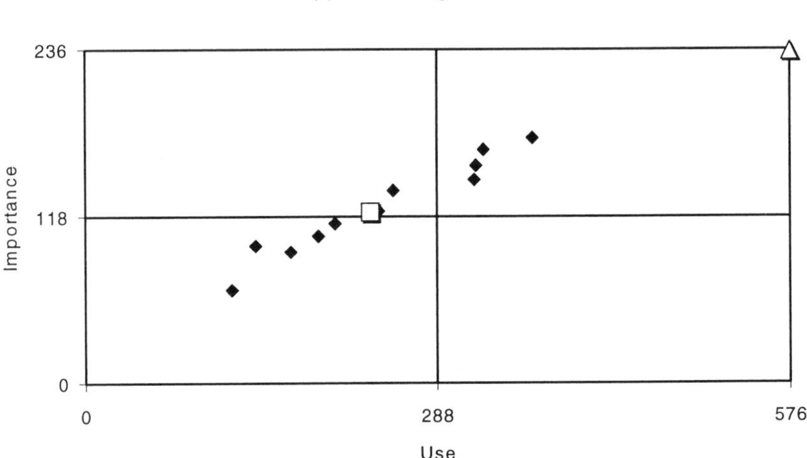

Figure 11.2 Supplier management. Industry average, □; out-of-sector comparison, △

Table 11.2 Supplier management: descriptions of practice

World-class best-practice company	Mid-range construction performance
Automotive components	
The organisation emerged from the bad old days of what was left of the UK automotive industry and has made enormous efforts to put that image behind it. Today, it is 100% reliant upon its IT infrastructure which extends forwards to customers and backwards to suppliers. That said, much of the organisation's success and current performance is built upon cultural values that emphasise investment in people and cooperation throughout the supply chain. Around half of the organisation's shares are owned by staff. A significant element of its work is the close management of its supplier network, which it relies upon to meet its commercial obligations. Indeed, it regards its position in the marketplace as determined by the effectiveness of its supplier management relationships and its IT. The organisation could not remain long in business if its IT infrastructure were to collapse. Since this is clearly a matter of vital concern, numerous safeguards are built in to its system's support to	Within supplier management, the organisation is primarily using systems for transaction management. The organisation has not yet fully determined the basis of measures needed, but realises the importance of doing so. The organisation has looked at cost savings in detail, but finds it more difficult to evaluate and quantify benefits. Today, the latter are seen as being very apparent and can easily become key to the organisation's normal working practices. Consequently, less importance is placed on the need to measure benefits. In terms of maintaining a competitive edge, the organisation has to change to stay ahead of the competition. The view is that it could not function efficiently without the use of IT. The use of purchase cards is becoming widespread, as is summary billing. Consolidated supply leads to sizeable reductions in the numbers of purchase orders being processed. The organisation is gathering metrics on numbers of suppliers and invoices and is monitoring the value of orders with individual suppliers. The average size of an order has increased greatly. IT is used extensively in planning for and monitoring performance. Overall, extensive use is made of IT at the level of transactions, but insufficient use is being made at the strategic level. The organisation is presently carrying out a full review of all processes and is looking at various ways in which IT can offer support and solutions, including the use of EDI. BACS is also being considered. The ability to be able to process invoice data on line may produce environmental savings, in terms of reduced paper consumption, and reduced head count. The organisation is

Contd

Table 11.2 *Contd*

World-class best-practice company	Mid-range construction performance
ensure that failure does not occur: if it did, the organisation would barely last a day or two. Given the organisation's origin in the automotive industry, it is perhaps not surprising that it owes much of its present transactional IT capability to systems developed many years ago. EDI is used extensively and is the main means for communicating with its suppliers. IT is being used to bring transaction costs down to zero as part of a focused assault on waste in the supply chain. Despite almost total reliance upon IT, the organisation stresses the necessity of involving all stakeholders in helping to find better, less wasteful and, therefore, more profitable ways of working.	looking at the use of IT in performance management, both with regard to its suppliers and in supporting the purchasing function. The intention is to use IT to facilitate increased performance monitoring, including the benchmarking of performance during projects. Supplier performance and supplier development are key to the future of the organisation. Data conferencing, particularly at the estimating stage of projects, is being considered. Process integration with suppliers and an IT support structure is planned. Data management is being addressed. Whilst the organisation has looked at improving the quality of data, there is still no clear guidance on who is responsible for managing this quality. This function is of key importance, particularly for a fully integrated systems approach. The organisation admits that it is not yet managing multi-owned data as efficiently as it should be. It views supplier management as a key process of the business. 75% of turnover is related to supplier management, with 80–90% of its work supplier related. On the data management front, the year 2000 issue is significant. The organisation is already requiring its suppliers to show preparedness for the year 2000 as part of the selection process.

11.4.2 Construction site processes

Interpretation of overall performance

Figure 11.3 shows a cluster of four companies close to the industry average. Three other pairs of companies are dispersed variously around this point, with one company significantly adrift. We might say that two, possibly three companies, not only consider the use of IT as important, they are also major users. These two or three companies may not have far to travel before they achieve best-

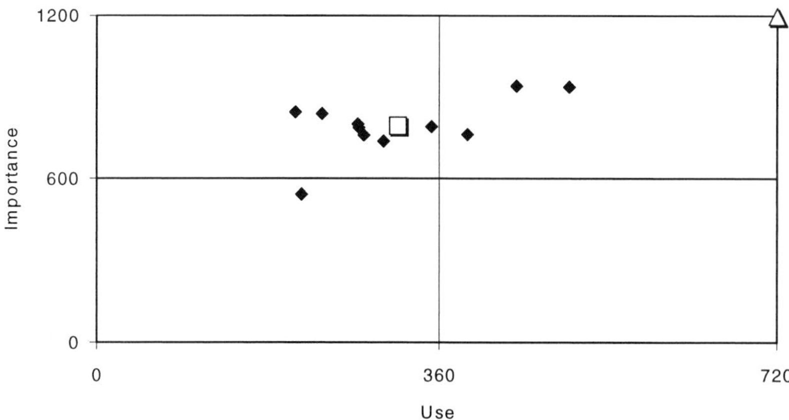

Figure 11.3 Construction site processes. Industry average, □; out-of-sector comparison, △

practice IT. The gap between them and the out-of-sector, best-in-class company is not so great that it could not be bridged by a concerted effort to invest in more and better IT. The majority of construction companies see the importance of using IT on their sites and do, in fact, use IT to a moderate extent. Two companies might be described as having a strong commitment to the idea of using IT, but have not made the investment. For one company, the position looks uninspiring. Not only does it fail to use IT to any significant extent, it does not see it as particularly important to the management of construction site processes.

Qualitative descriptive comparisons

Table 11.3 compares best-in-class construction site processes with mid-range performance.

11.4.3 Briefing and design

Interpretation of overall performance

When overall performance is averaged for each organisation, according to the stages that each covers, the result, shown in Figure 11.4, is more widely dispersed positions. A line of best fit would again be straight, indicating that IT is used, more or less, to the extent that it is considered important. The performance of client-

Table 11.3 Construction site processes: descriptions of practice

World-class best-practice company	Mid-range construction performance
Nuclear decommissioning The company has developed an integrated approach to the management of construction site processes. This is based upon a suite of over 300 automation tools, otherwise known as standard application programs or modules. The objectives of this integration are to: (1) reduce total installed costs; (2) shorten project programmes; and (3) support globalisation by establishing electronic data linkages that will bring together office, sites, customers and suppliers irrespective of their physical location. The main functions of this integrated systems approach are to: • provide services and IT tools that automate and integrate project work processes, including concept design and estimating, detail design, materials management, construction and project management; • increase productivity on projects by reducing the time spent on repetitive tasks, reducing or eliminating redundant data entry and increasing accessibility to data required for different processes; • provide current information such that decision-makers can see data from multiple sources within a single context; and • include provisions for EDI (electronic data interchange) with customers and suppliers and linkages to third party software.	By sharing data electronically, we have helped to reduce the distance between head office and site. The company operates a policy where the project manager can work from head office for 2–3 days per week whilst still exercising effective project control. The project manager is responsible for the design and construction with an empowered multi-discipline team. Empowerment has allowed the team to choose and experiment with software and systems when they see fit. The company uses a 3D CAD modelling system as a central production tool in support of site activities. In the future the company will adopt systems for document management, video conferencing, electronic timesheets and procurement and will address issues of integration.

led organisations is strong, as in the overall picture, with these organisations joined by a significant number of those that are designer-led. Again, one organisation – the same as for overall performance – stands apart. However, actual use is not so far advanced that it is performing out of reach of the rest. For a few organisations, there is much to be done to improve performance.

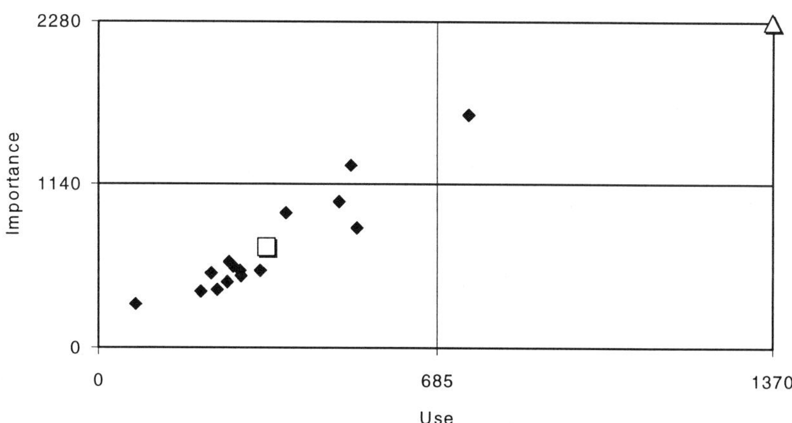

Figure 11.4 Briefing and design. Industry average, □; out-of-sector comparison, △

Qualitative descriptive comparisons

Table 11.4 compares best-in-class briefing and design with mid-range construction performance.

Table 11.4 Briefing and design: descriptions of practice

World-class best-practice company	Mid-range construction performance
Automobile manufacture A key factor in the success of the company, and in securing its future, has been the use of CAD, knowledge-based engineering (KBE) and other state-of-the-art IT. They can do things today – in design especially – which they could have never done before. But that position is not unique to the company. In a global marketplace, their competitors have not been slow to adopt and adapt newer forms of IT. CAD is routine and KBE will one day achieve the same status. IT is a key enabler of engineering and provides tools which have become a fundamental part of the organisation. The company's strength is derived through its relatively compact size which enables it to compete in a global marketplace. Success is also due to its ability to be inventive and innovative. Information	The aim is to maximise the appropriate use of IT within the business constraints and to take advantage of what is available. IT use will grow through improved availability and accessibility. Network-based communications will be improved, enabling better access to corporate information.

Contd

Table 11.4 *Contd*

World-class best-practice company	Mid-range construction performance
is a vital commodity, as indeed it is in many organisations. In coping with engineering design, the performance of individual components has to be fully understood. Presently, much information of this kind is paper-based, but the company is moving toward on-line documentation which will include design procedures. This is being developed and will be implemented as part of a company-wide intranet. The company claims substantial savings in product definition, design and prototyping by the use of KBE in particular. This has resulted in marked reductions in engineering lead-times, thereby enabling it to bring its new models to market that much sooner. In summary, this world-class, best-practice company uses IT to: • reduce the time taken to engineer a basic structure, in some cases, from months to days; • redesign components and parts in a few hours, instead of days; • give more time to design engineers to investigate further options reflecting better trade-offs; • integrate design and manufacturing knowledge into a single product model that enables engineering issues to be resolved concurrently; and • refine the rules held by KBE models so that information and expertise which was once diffused – in manuals, data-sheets and in people's heads – is now captured in a single representation.	3D and project modelling will be used more and more to support a *virtual build* approach and will rely on object-oriented techniques. Document management is an area for further development, alongside that of improved project management tools, better access to management information, CD-ROM (for information dissemination and storage) and video conferencing. Greater use will be made of portable forms of IT.

11.4.4 Facilities management

Interpretation of overall performance

Figure 11.5 shows a consolidation of scores, when compared with overall performance. The variability in performance is less dispersed with a greater clustering of scores, pointing to the need for significant process improvement across the board. There is one rather isolated performer whose IT would seem to lag behind the rest by a significant margin. For the leading cluster, there is a discernible move away from using IT only to the extent that it is

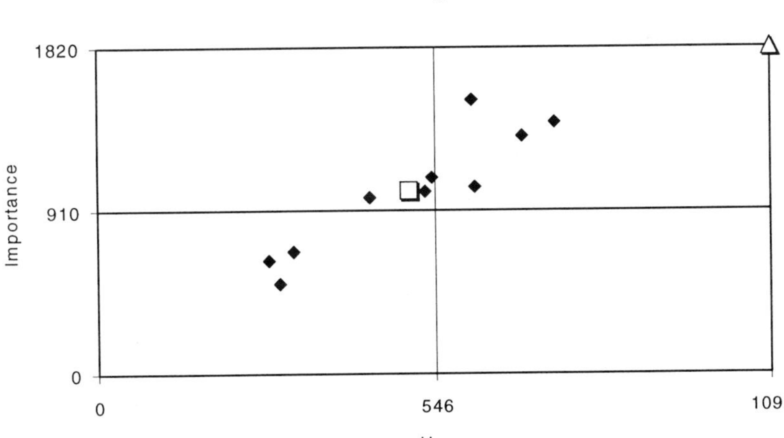

Figure 11.5 Facilities management. Industry average, □; out-of-sector comparison, △

considered important. The inference is perhaps that IT use may, in some cases, not be considered as absolutely essential to success, but it is used nonetheless. This trend is, however, only marginally evident in Figure 11.5 and it would not do to place too much emphasis on this point.

Qualitative descriptive comparisons

Table 11.5 compares best-in-class facilities management with mid-range construction performance.

Table 11.5 Facilities management: descriptions of practice

World-class best-practice company	Mid-range construction performance
Real estate management The organisation takes total facilities management, when required, for all the buildings under its management and has to meet strict performance criteria – everything is results-driven with management constantly on the look-out to add value. There is a very clear structure to the organisation's operations. Instead of trying to pull together disparate packages and systems, the decision was taken to link all applications through Microsoft SQL Server – a high-performance database management system	This major owner of buildings throughout the UK covers all areas of facilities management, though is less active in environmental management at this time. Its annual spending on operations is in the region of £25 million. Some of its

Contd

Table 11.5 *Contd*

World-class best-practice company	Mid-range construction performance
running under Windows NT – to provide the highest possible degree of integration. A distributed, client-server computing model has been adopted. Financial management, real estate functions, engineering and supporting administration are linked to provide a high level of automated functionality. As far as possible, all data are keyed once. Applications then use those data to provide management with the information they require to deliver the best possible service to the customer. This inevitably means pulling data from several different parts of the system. Development of IT is a constant feature, but there is no experimentation on live projects. One of the principal aims of IT is to focus on routines that are a chore for people and either to eliminate them or substantially reduce the time needed. The organisation makes good use of standard, low-cost, off-the-shelf software and e-mail. There are essentially two types of IT applications when taken in the context of computer-based systems. First, there are those which have been developed specifically for the organisation to provide it with a competitive advantage and which contain knowledge about its operations at the strategic level. Second, there are those which are essentially open systems for which it is happy to share in development work with others. At least 10% of an individual's time is spent in training of one kind or another. In the future, the organisation will seek a better balance between IT capabilities and those of the user. The belief is that within 4–5 years, IT will have been transformed in ways which are difficult to predict now. Foreseen changes are the availability of software which fits the user's profile and self-learning software to correct an individual's errors and to extend his or her capabilities.	facilities management is outsourced, for instance domestic services which include cleaning, pest control and laundry items. Future plans are aimed at greater integration between systems, or at least compatibility. IT will be used for customers to generate their own requests for facilities management support without having to pick up the telephone. IT will also enhance the effectiveness of services provision. A well developed IT infrastructure will include the supply of services such as air tickets and other forms of ticketing. This implies stronger links between departments for the expansion of services. By broadening the meaning of facilities management, it will be the first port of call for all internal requirements. Management information systems need refining. The information could minimise space utilisation by enabling hot-desking and teleworking.

11.4.5 Cost estimating and bidding

Interpretation of overall performance

Figure 11.6 shows a wide distribution in the performance of the companies, with a cluster of six demonstrating significant use of IT that is underscored by its importance. However, the gap between them and best practice is wide and suggests that, for some sub-processes, far greater penetration of IT is necessary. For three companies and one in particular, the gap between their use of IT and those at the fore is considerable – more than the gap between the leaders within industry and best practice. Despite this picture, all companies regard the use of IT in cost estimating and bidding as important. The difference between the companies is that some have done far more than others.

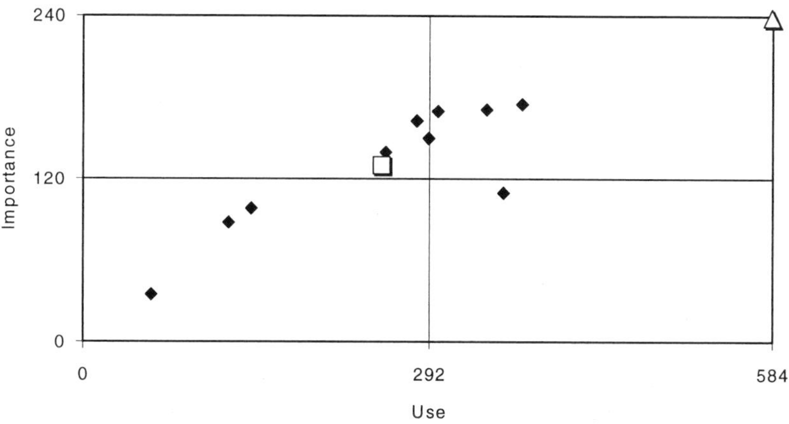

Figure 11.6 Cost estimating and bidding. Industry average, □; out-of-sector comparison, △

Qualitative descriptive comparisons

Table 11.6 compares best-in-class cost estimating and bidding with mid-range construction performance.

11.4.6 Cost and change management

Interpretation of overall performance

Figure 11.7 shows a wide distribution of performance amongst the organisations across the breadth of this process. The gap between

Table 11.6 Cost estimating and bidding: descriptions of practice

World-class best-practice company	Mid-range construction performance
Engineering production This best-practice organisation has invested heavily in its IT infrastructure. All computer workstations – based on various kinds of PC – are linked by LANs and run under Windows NT, using TCP/IP for communication. Servers are located strategically to prevent mishap in the event of fire or other event and are backed up continuously as well as being mirrored. Groupware is used to provide the basis of an intranet that enables information to be exchanged internally and documents to be managed in a way that minimises hard copy. Documents such as correspondence and invoices are scanned upon receipt and individuals notified of their availability. Microsoft's SQL Server is used as a centralised database from where applications, with respect to various technical and administrative functions, obtain and deposit their data. When new applications are written they have to be integrated within the existing arrangement. Links externally are achieved by ISDN, allowing Internet access to a remote server where the organisation's web site is hosted. Suppliers are notified of bid situations by e-mail, but must go to the web site to download the necessary documents before they can begin the process of cost estimating. Quotations are always submitted electronically, but must be followed by hard copy if they are subsequently accepted. Likewise, several of the organisation's customers now require bids to be delivered electronically, with these too followed by hard copy. The short periods allowed for bidding mean that more time can be spent on getting the price right than worrying about the physical delivery of a document by a deadline. Even so, the organisation always checks	The organisation has a LAN in the estimating, accounting, personnel and marketing sections operating under Novell and based on Ethernet. The three main offices are connected by kilostream line, with connectivity to satellite offices from dial-up links via the central server. Estimates, material analyses and other associated documentation are often transferred electronically between offices. Some use is made of e-mail attachments to and from the design team, particularly where design and build projects are involved. The extent of IT used is determined by the ways in which the organisation receives data. Clients more usually request software compatibility rather than prescribing specific systems. The primary advantage of client IT requirements is to maintain connectivity, to be able to move data quickly and to avoid the double entry of data. Scanning is limited to accounts, especially invoices. Scanning has been attempted for other documents, but discontinued. For the time

Contd

Table 11.6 *Contd*

World-class best-practice company	Mid-range construction performance
by telephone that its quotations have been received when it has not received a reply or acknowledgement within a prescribed time. Pricing is a semi-automated routine, based on well-defined procedures, that requires intervention only when the final price is being considered. Manual checks ensure that data are not missing, for example, making sure that all quotations have been properly considered and that all items have been priced. The organisation sees various opportunities for streamlining its IT, especially in how it communicates with its suppliers. The most obvious development is for the web site to be brought in house and for an open, high-speed line to be installed. This will allow suppliers to look more closely at the organisation's needs and not to have to be prompted whenever a bid is underway. The organisation recognises that more effort and money must be put into ensuring that its suppliers and customers are able to communicate with it as effectively as possible. This means it will undertake a limited amount of IT development with its suppliers where it feels this would be repaid through a 'less hassle' arrangement. Upgrading of many of the PCs is a challenge for the organisation, because it recognises this is going to be a fairly regular event and a costly one too.	being, cost of storage and ease of access are still issues, with difficulties encountered in scanning and retrieving documents. In the longer term, the organisation is sure that actually scanning will be the more cost-effective alternative to traditional means. The organisation is in the process of introducing a full WAN, which should be operational by the date of publication of this report. It is looking at more integrated solutions, particularly for post contract works, and has implemented a bespoke system on a pilot scheme. In the short term, however, the organisation will continue to use its existing packages and procurement systems. What is required across the industry is one integrated package that starts with marketing and ends with the final account.

the best performers and the weakest is roughly equivalent to the gap between the former and best practice. For those whose performance is amongst the weakest, there is some way to go before they are on a level with the best, by which time the latter will probably have moved closer to best practice. Disappointingly, more than half of the organisations demonstrate only a moderate use of IT. Additionally, one or two organisations also attach modest importance to their use of IT.

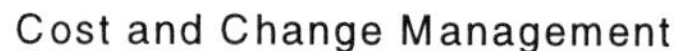

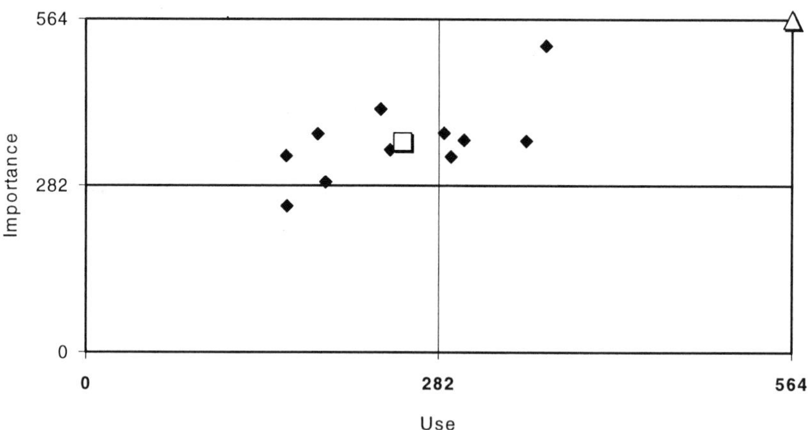

Figure 11.7 Cost and change management. Industry average, □; out-of-sector comparison, △

Qualitative descriptive comparisons

Table 11.7 Cost and change management: descriptions of practice

World-class best-practice company	Mid-range construction performance
Manufacturing generally An approach that supports the electronic capture and manipulation of data to add value to the core business is central to a best-practice view of any process. For cost and change management, this can be extended to linkages with other parts of the business. A best-practice approach is therefore implicitly about capturing data about the client's needs, the project's objectives and its constraints, making reference to design models and their associated costs in order to guide the decision-making of the project team. Cost and change management has to work ahead of events, that is, it must enable project teams to anticipate changes and interpret the implications in a way that	The organisation is the project management division of a major cost consultant, specialising in project management and support, design and cost management and risk decision support. It is typically involved in projects of varying size from initial feasibility through to final account, covering building construction, rail, road and process engineering. Most information within the organisation is held electronically and also transmitted electronically via e-mail attachments or the company intranet. Frequent inter-organisational use is made of e-mail and attachments for text documents, spreadsheets,

Contd

Table 11.7 *Contd*

World-class best-practice company	Mid-range construction performance
does not compromise the project's objectives. The process is dynamic and continuous throughout the life of a project. The aim is to have the means to be able to report at any time on the current financial state of a project, its design and construction implications. In IT terms this does not require much sophistication. The main challenge is a procedure for ensuring that deviations from a plan or other agreed action are checked. Having access to necessary data is paramount and so a best-practice IT approach to cost and change management is one that can immediately show the likely or actual impact of a deviation in the design. Furthermore, the IT must provide seamless electronic support and interconnection with the function that it is mainly intended to support, in other words, design and its realisation through construction. Links to CAD will be, therefore, an integral part of an IT system that both supports the exchange of data and tracks the decisions that bring about such exchanges. Inevitably, this will mean IT tools for database management, adopting a client-server model, where reports can be generated to suit the needs of the different members of the project team. In this way, no one can be left out of the picture other than by intention.	database and programme information. Occasionally, disks are posted, but e-mail is now more likely to be used. Some use is made of scanners, often on sites, and use is also made of microfiche for archiving drawings. Permanent connections are maintained to other divisions and overseas offices. Since Internet access is available on every desktop, the organisation considers little additional benefit would accrue from separate on-line information systems. The organisation is looking at issuing information on CD, particularly drawings, *as-built* information and O&M manuals. They are in the process of having a knowledge base put together to close the loop between the proposal and project execution phases, enabling the organisation to store information for subsequent use. The valuation and certification modules for an existing project management system are presently under development. The intention is to link the change control module in it with the cost planning side, in order to improve cost reporting, that is, to bring about greater integration.

11.5 Implications for the firms

In general, the distance between the best construction performers and the worst is significant, though the gap between the best and the associated out-of-sector comparison is, for most of the studies, reassuringly smaller. Useful lessons can be learned from best-practice IT as introduced by these out-of-sector comparisons. IT is vital to their mission-critical business processes and is reflected in a high degree of integration of processes and activities within them. This integrated approach is no accident, but caused by a well intentioned strategy for raising efficiency, lowering operating costs and delivering customer satisfaction.

Success in the long run, in terms of market share captured by individual firms, will be linked to their ability to develop and apply IT-based systems that will contribute significantly to reducing the cost of providing services and raising value for money for the customer whilst, at the same time, enhancing the core business.

11.6 Implications for the wider industry

Many lessons can be learned from the findings of this series of benchmarking studies and the best-practice IT out-of-sector comparisons. They can be thought of as factors that are key to the successful implementation and exploitation of IT in support of mission-critical business processes.

- Firms generally must re-examine their underlying business processes and ensure that they are aligned to delivering customer satisfaction and value for money.
- A strategy that elevates the importance of IT in supporting mission-critical business processes is required of all firms and this must include a clear statement of where the firm is aiming its sights.
- Successful implementation of IT within firms is reliant upon an IT infrastructure of the kind that will allow for open communications based on industry standard protocols. This has to become the top priority, since some IT supported activities are impossible to improve without the appropriate investment in infrastructure.
- Integration is the key to success in terms of process improvement. It can be achieved in various ways, but the consistent theme is by the application of industry standard systems that enable changes in operating procedures to be introduced as quickly and as efficiently as possible.

11.7 Research and development priorities

All benchmarking studies in this series have measured two factors:

- how important IT is for a particular process; and
- how effectively IT is being used to support that process.

By depicting these two measures within a grid, as shown in Figure 11.1, it is possible to classify a benchmark as shown in Figure 11.8.

Figure 11.8
Classification of benchmarks

Important area where IT is poorly used	Effective current use of IT
Non-priority area for IT at present	High, ineffective use of IT

This classification can now be used as a pointer for research and development priorities (Figure 11.9).

If the results of this benchmarking study are now superimposed onto Figure 11.9, it is possible to categorise the overall performance of the seven studies, as shown in Figure 11.10.

The findings in Figure 11.10 suggest that a selective approach to

Figure 11.9
Identification of R&D priorities

Short-term research priorities	System development priorities
Possible long-term research priorities	Non-priority area

Figure 11.10 Priorities for effective IT application

Briefing and design Facilities management	Cost estimating and bidding Cost and change management
Supplier management Construction project site management Integrated project information	

research and development is required. For example, briefing and design has reached the point, in terms of IT support, where a serious push is needed now to enable it to create a seamless process from predesign to the production of detailed drawings. For other processes, for instance construction site project management, much has to be done to create a critical mass of IT support that will enable data to be integrated and shared between applications. In all cases, there is ample evidence provided in the out-of-sector comparisons to show how real improvement is possible.

11.7 Summary

The studies provide reliable snapshots of current applications of IT in the UK construction and engineering industry. Repeating them at some time would reveal where progress had been made and where attention is still required. Benchmarking is, after all, a tool for continuous improvement. Given the findings of the studies, best practice IT is within the reach of some firms within the industry. Their progress towards that goal can be measured periodically. Even so, they must clarify their business objectives, adopt clear policies and commit themselves to implementation of IT on a major scale. Those that do so will be amply repaid, in much the same way as the out-of-sector exemplars of best practice.

12 Benchmarking project information integration

Brian Atkin, Paul Davies, Anne Marie Dubois, John Gravett, Adina Jägbeck, David Smith and Stephen Walker

"The information flow between project participants is the key issue in bringing about integration between the different project life-cycle stages."
– *Italian architect*

The benchmarks of IT use to support discrete construction processes, described in Chapter 11, give a very detailed insight into relative performance within construction. The larger goal with our use of IT in the sector is to achieve integrated information management throughout projects. Benchmarks of leading information integration practice across six major European projects again reveal a high variability in performance. Situations of greater information integration tend to occur when parties are physically remote from each other, when there is a single-point design and construction responsibility and where the client encourages integration.

- The primary technical aim of strategic IT use on projects is integrated information management and exchange between all parties at all stages of the life cycle.
- By visiting and interviewing multiple participants on six major EU projects, the processes undertaken and the method of information exchange have been documented.
- Findings from these studies are that highly variable levels of integration are achieved in practice on major projects.

"We are running a dozen different systems at the moment. Some driven on the left, some on the right, some in the middle."
– *Contracts manager of UK contractor on EU joint venture project*

12.1 Introduction

This chapter describes some further benchmarking undertaken as part of the work of the Construct IT Centre of Excellence. This benchmarking is of project information integration undertaken as part of the EU SCENIC project. SCENIC (Support Centres Network for IT in Construction) is an ESPRIT-funded project to create a best-practice network of construction IT initiatives in different European countries. It performs a central role in bringing together different players from within these fields to create an organisation to help improve the industry's performance through heightened awareness of the use of IT and a process of technology transfer.

The SCENIC project:

- bridges the gap between IT and the construction industry;
- is responsible for disseminating information on available and emerging technological innovations; and
- offers specific assistance to a large number of small and medium sized enterprises within the industry.

These ambitions are being achieved through three major action initiatives:

- identification of needs and best practice;
- survey of ongoing research and technology development; and
- dissemination and technology transfer.

The last of these ambitions is partly realised through the work presented in this chapter, that of a study of comparative IT use in integrating project information in three European countries. SCENIC representatives in France, Sweden and the UK were each asked to put forward two projects so that the use of IT could be compared. The findings from these six projects provide valuable insights for construction professionals across Europe. Benchmarking is both the theme and the tool used in that work.

12.2 Background

The primary benefits from benchmarking use and performance in IT are in providing an objective assessment for those organisations forming the study group. The publication of the results of the study also means that information about the use of IT is disseminated to a far greater audience. In this way, valuable lessons can be learned by a wider section of the European construction industry than would hitherto have been possible. However, the names of the participants and their projects are not disclosed in this chapter as that would breach a basic tenet of this kind of benchmarking, that of anonymity and confidentiality. The European dimension of this work is the comparison of different projects in different member states. That said, there are many similarities, not least their approach to the design and construction process. Whilst procurement arrangements vary and the sequencing of activities may well differ, design must precede production and assembly and other aspects must take their logical place in the order of works.

The focus for this benchmarking study is integrated project information. The rationale for its choice was that the use of IT to support the exchange of information, amongst project participants, is significantly more important than individual applications of IT. Study at the project level also provides a common basis for comparison as well as showing how IT is being applied at a level that is fundamental to the work of the industry and one of the levels for strategic management we considered in Part A. Chapter 11 illustrated benchmarks of IT support to discrete project processes. This chapter illustrates how IT allows project information to be integrated throughout the process.

12.3 Scope

Integrated project information is a measure of the extent to which information in support of the total building life cycle - taken from inception through to occupation and re-use - is integrated. It is about communication between participants (often referred to as actors) that is both electronic and seamless across all stages in the process. The unit of analysis is, therefore, the project not the participants' organisations.

All relevant stages in the building life cycle were considered and most were analysed in the study. Comparison was made on the basis of this generic model of the building life cycle, taking account of different procurement strategies, and the extent to which information flow was both electronic and seamless.

All the case study projects were well advanced in their construction and each was believed to reflect good use of IT. The study team did not seek the best projects from an IT perspective, rather it chose those that were indicative of current good practice in IT. Through comparison it has been possible to show how each project used IT to support its underlying processes.

12.4 Generic stages in the building life cycle

To allow a level of commonality between the studies conducted, a generic life-cycle stage definition was set at the start of the work. This was then adapted according to local conditions by the study team. The generic stages were:

(1) inception;
(2) briefing;
(3) feasibility;
(4) concept design;
(5) scheme or outline design;
(6) detail design;
(7) tender documentation;
(8) estimating and tendering;
(9) evaluation of tender;
(10) off-site fabrication/prefabrication;
(11) delivery/logistics;
(12) production/assembly;
(13) testing, commissioning and hand-over;
(14) operations/facilities management; and
(15) re-use/demolition.

The benchmarks that follow typically applied to stages 1–13 of the above list.

12.5 Method

At the outset, it was agreed that all observed information flows should be regarded as necessary to the projects in question. This might seem an obtuse point, but to have attempted an assessment of every piece of information was considered outside the scope of this project. Participants were interviewed on the project sites or in their offices in order to obtain as complete a picture as possible of the information flows, their frequency and the kind of IT being used. Each project was visited, on average for two days: first, to get the general impression of information flows and IT use, and second, to make detailed notes. Further days were spent in preparing and validating the information models.

Interviews began by considering the extent to which the list of generic stages applied to the project. The list was accompanied by an IDEF0 diagram (see Figure 12.1) and a questionnaire for capturing details of the information flows. These interviews were usually held with the project manager, followed by each of the key members of the project team until a sufficient level of detail had been captured to support analysis.

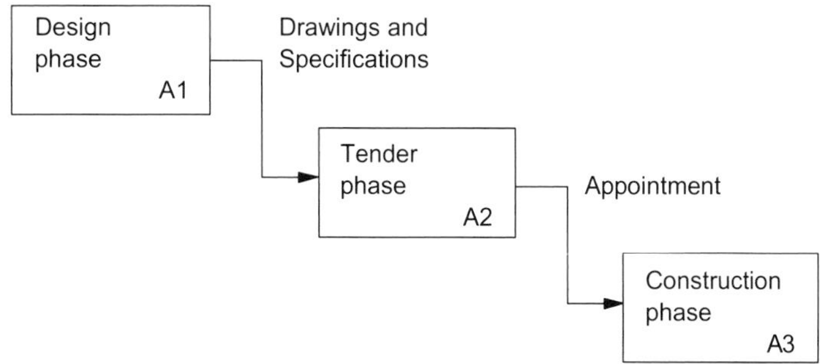

Figure 12.1 IDEF0 diagram of process and information flow

In terms of analysis, electronic transfer between project team members was the principal basis for measuring the extent to which information was integrated. The extent of electronic transfer of necessary information is, therefore, the measure of the degree of integration on the project. The result is an index number which can be used to describe the degree of integration achieved on these projects. An index number approaching 1.0 would indicate a very high degree of integration of information; conversely, numbers approaching 0 would indicate very little integration.

12.6 Degrees of integration of project information

The six projects studied displayed very different levels of project information integration. On a scale of 0 to 1, the range within these six projects varied from 0.22 to 0.82. This demonstrates that even from a very small sample, there is great variability in the extent to which information is integrated on major European projects. The reasons for, and implications of, this, are the findings from this chapter.

Project	Index	Grading
A	0.64	Strong degree of integration
B	0.40	Moderate degree of integration
C	0.82	High degree of integration
D	0.51	Moderate degree of integration
E	0.22	Low degree of integration
F	0.56	Moderate to strong degree of integration

12.7 Main findings

(1) IT is being used to counter the difficulties of disintegrated project-based work in this fragmented industry, by supporting the electronic exchange of project information.

(2) IT is being used to support innovation in methods of procurement and in technology procurement. The highest degree of project information integration is where there is a single source of responsibility for the project.

(3) Integrating project information is better supported by IT when there is a physical distance between the participants, for instance when they are located in different regions. Efforts to effect better communications are helped by the use of IT.

(4) IT use is generally lower where project teams are co-located. This working arrangement has benefits for team-building, but tends to militate against coordination of information and can succeed in building paper bureaucracies.

(5) IT makes it possible to think seriously about a European construction industry and to underscore, with confidence, the role that it must play in bringing about an efficient and competitive industry.

(6) IT enables more firms – especially SMEs – to enter markets in member states that are not their traditional territory and enables their efforts to be integrated with the rest of the project team, allowing them to work on an equal footing.

(7) IT use would be further enhanced if models of the kind prepared in this study were developed and made widely available – some form of generic project information model (or process map), suitable for adaption, would provide a useful basis for defining the operating parameters of projects.

The case studies

Project A

Considerable use was made of IT to support design activity on this major project for a government department. Key members of the project team were interviewed, including the project manager, architect, main contractor and selected specialists.

All designers and specialist contractors in this multinational project team were required to work with CAD from the outset so that they could exchange their working files with the design team. Early in the project, preliminary drawings were placed on the project server, which was accessible to all actors having an interest in the design. As design was progressed by the various designers and specialists, work was added as layers over the original, preliminary drawings. Validation of the design, including checks for consistency and clashes, was undertaken at working meetings and decisions recorded on the files held on the project server. Whilst this use of IT and the practice it supports is hardly new, it is the rigour with which this work was carried out that makes such use worth reporting here.

In broad terms, this project demonstrates a significant degree of vertical electronic integration of activities within the design process. The process maps produced for the project reveal a complex network of activities that required careful and close management. IT proved especially beneficial in this work, by allowing the efforts of individual actors to be aligned so that consistency was maintained throughout.

Management of the construction phase was supported by IT to the extent that it was used to monitor and evaluate progress of the project as the basis for determining payments to contractors. Again, the use of a central server for the project has helped to reduce the amount of re-keying and scope for error – this latter aspect is especially important for safeguarding the financial management of this complex project, as well as ensuring that contractors and suppliers are paid properly. Apart from the widespread adoption of CAD and the need for resident expertise in the use of the system, there was also an engineering and technical coordinator and a software manager.

Overall, this project exemplifies how IT can be used to impose a beneficial discipline on a project team whose origins and experiences are diverse. This is especially true of the use of CAD and the procedures surrounding its application. Success in this project was derived from the close cooperation of the project team members and their willingness to see IT as a means for helping to integrate their efforts as much as in providing a platform for the exchange of project information.

Analysis of the information flows on the project results in an index number of 0.64, reflecting a strong degree of integration.

Project B

This large-scale, social housing project followed a traditional pattern of procurement that was as demanding as it was predictable. IT was used to support some innovation in a design and construction process that was geared to a tight programme for the project's completion. IT was used in coordinating the work of the various actors in the team by enabling data to be exchanged electronically. Key members of the project team were visited, including the project manager, architect, cost manager, and health and safety manager. Their contributions provided a comprehensive account of the design and construction process, the information flows, other actors involved and use of control mechanisms. In the first instance, the process maps brought together the different perspectives of these various actors and the information used by them in their work. Subsequently, tables were compiled, providing descriptions of the IT tools used: comments were sought as to the opportunities for optimising the benefits from IT on this and future projects. What was produced was a sound definition of the process and the part that IT played in it.

The project was complicated by its size, creating a major information management task for the project team. IT was used in feasibility and in programming the construction works. Many of the tools were generic, but no less effective. For instance, use was made of spreadsheets, project planning and scheduling tools. Earlier use of IT during concept design and, before that, briefing was however limited. IT was used later in detail design and contract administration to good effect. In many respects, this project's more effective use of IT was constrained by a statutory planning and approvals process that largely dictated the pace of the project.

The main finding was of IT being used to draw together the working practices of project team members. A more developed use of IT would have helped in integrating a greater proportion of the information used by team members. That said, the process maps prepared in the course of the study served a valuable purpose in showing where IT need was the greatest in managing the integration of information.

Analysis of the information flows on the project results in an index number of 0.40, reflecting a moderate degree of integration.

Project C

Innovations in practice, as well as in the use of IT, are reflected in this case study, the design and construction of a medium-sized, speculative housing project. A striking feature of this project is that of a single source of responsibility under the control of a developer-builder. The adoption of an innovative procurement method took the place of practices that were long considered in need of change. A good example of this was where management of the supply chain had to be completely overhauled. Suppliers became partners in the process and, as such, were given electronic access to the project database using ISDN. This innovation enabled the project team to eliminate many non-value added activities that otherwise characterise much of traditional procurement. Yet, none of this happened overnight. Considerable R&D was necessary, but the benefits were clearly visible both in the pace of the project and in the financial return on investment.

This project exemplifies the concept of integration through a re-engineered process that used significant IT to achieve lower production costs against tighter timescales. Indeed, the project team was able to boast substantial reductions in programme time when compared with more traditional procurement, including their own previous work.

In overall terms, this developer-builder has claimed a 50% reduction in time spent on design today, compared with earlier projects. Particular gains arose from the ability to make changes to the design once only and to communicate them rapidly to those affected. CAD is a key technology for the developer-builder and one that is being used to build object-oriented models embodying information about the process. The intention is that these can ultimately be evolved into product models serving the wider needs of the project team, manufacturers, suppliers, the client and end-users.

During the project's design, checks for consistency in detailing and clash avoidance – especially in the area of engineering services – also meant that prefabricated components could be manufactured in the knowledge that few problems would arise on site. This latter aspect was important since the time spent on site had been compressed to the point where there could be little tolerance for delays caused by errors and inconsistencies. Bringing manufacturers and major suppliers into the project during the earliest of stages helped to coordinate work to the extent that the goal of zero errors in component delivery and assembly is now within reach. Comparison of the use of IT against the earlier manual process showed that quality, reliability, accessibility and re-usability of information were significantly better.

In many respects, this project has moved a long way towards the ideal of making construction more like a manufacturing process. By applying IT to a re-engineered process, the developer-builder has been able to make the kind of progress in integrating project information that would be extremely difficult under more traditional procurement. Despite the enormous gains on this project, the developer-builder is intent on extracting further gains as part of a culture of continuous improvement. In this regard, he is willing to engage in discussion with others and to share knowledge that will raise performance higher still.

Analysis of the information flows on the project results in an index number of 0.82, reflecting a high degree of integration.

Project D

For IT to play its part in delivering a project on time, within budget and to the required quality does not mean that expensive or sophisticated tools must be used. In some cases, as this project demonstrates, the use of low-cost, widely available IT can, when applied thoughtfully, bring useful gains from more focused working. This project, the design and construction of individual houses, involved the prefabrication of major components intended for rapid assembly on site. The SME behind this project, acting as a developer-builder, shows how it is possible to manage projects of a size not normally associated with small enterprises. Often such enterprises will operate with little or no IT to support this kind of work, with the result that they employ human resources at a level well above that found in this case study project. IT can extend the capabilities of even the smallest of enterprises, if used in a similar way to that found here. This means that smaller enterprises can compete with larger enterprises when they are supported by an appropriate level of IT. The level of cost associated with the IT found on this project is not excessive or disproportionate to the enterprise's size. Scaling up the IT to the level of a medium to large size enterprise would, however, represent an enormous investment for those enterprises and one that is unlikely to be found in practice.

For the case study project, CAD was the core technology through which design, fabrication, cost estimation and schedules were either generated automatically or created by linking to other software packages such as spreadsheets and other office automation tools. Consistency and integrity were assured by avoiding re-keying and re-formatting data, and using a single representation of the design. Savings in time and resource were also significant and there was less likelihood that mistakes would occur in communicating with manufacturers, suppliers and customers. Comparison of the present use of IT against a process that was previously manually based showed that quality and reliability of information were better.

Analysis of the information flows on the project results in an index number of 0.51, reflecting a moderate degree of integration.

Project E

Major projects can create many challenges. Complexity and an innovative form of procurement were just two challenges facing the project team on this large building project as part of a public–private partnership. Add to this a diverse and culturally varied project team and one has the makings of a very demanding project. Perhaps for these reasons, effort was concentrated more on integrating people into effective teams than on a high level of IT use. Whilst IT was present in many forms – desktop PCs, e-mail and database management – there was little linking and sharing of information. Nevertheless, the project was viewed as a success, against a tight timescale, strict budgetary control and a demanding client. The nature of the project, in particular its intended use, meant that security was an important factor weighing on the project team. The imposition of rigorous procedures, designed to ensure confidentiality and security on what was a politically sensitive project, meant that IT presented too great a risk for the project team.

Limited electronic exchange did take place between various parties, for instance amongst the designers and with the client body, but even these did not amount to a significant attempt to integrate information on a project-wide basis. The demands of the project were probably thought to be enough without having to cope with the uncertainty of IT. Even so, the opportunity to enforce better communications between the various actors through, for example, project-wide e-mail and drawing exchanges was not passed up. Project team members were acutely aware of the penalty for failure on the project and so anything that was unproven was unlikely to be adopted. Given that the project was the first to bring the consortium together, there was arguably no obvious basis upon which to implement an information management strategy.

IT was used in many different ways, but most were in supporting the disparate working practices of the various actors. As a means for exchanging information, IT was rejected for the reasons given above. The co-location of project team members helped team building, but did not necessarily make for an efficient process. Meeting deadlines and working within budgets are not conclusive proof of an efficient and cost-effective process. Concentrating attention through the actions of project personnel may not represent the most efficient use of human resources, but it can get the job done. The fundamental issue is therefore one of whether or not IT is necessary for integrating the project team. On the face of it, there is no evidence to support this contention. However, neither is there evidence that efforts to integrate project information would have compromised time, quality or security.

Analysis of the information flows on the project results in an index number of 0.22, reflecting a low degree of integration.

Project F

When a project team is dispersed geographically, IT can be used to bring the team members closer together by allowing them to communicate and exchange information. On this project, a major headquarters building for a financial services company, IT use was most pronounced in its support for the traditional roles of the various actors. However, this use was modified to encourage information to be exchanged between project team members with the minimum of re-keying or re-formatting of data. The project was based in one city where there was a substantial site presence of personnel, but several project team members were established in, and working from, offices in other locations. This meant that more effort had to be put into supporting the exchange of information between the various actors whilst allowing them to work on their respective parts of the project. This posed a particular challenge for the project manager since a good deal of his time was spent in coordinating the work of others. This burden could well have been alleviated by more support from IT.

In terms of innovative uses of IT, there is little to report. CAD, project planning and scheduling and e-mail, together with office automation tools represented the core IT tools on the project. Little other specialist IT was deployed, but effort was put into facilitating data exchange with generally satisfactory results.

Analysis of the information flows on the project results in an index number of 0.56, reflecting a moderate to strong degree of integration.

12.8 Summary

Despite the intensive use of IT in some of the case study projects, not one of them had an explicit plan for IT use at the level of detail adopted by the research team for its analysis. In the course of interviews and in subsequent feedback sessions, it became readily apparent that little thought had been given to the need for mapping the information process in the way that the research team had adopted in its work. With very few exceptions, project team members saw the benefits of having an explicit statement of processes and activities and the information flows between them. The view of the research team is that a serious attempt to produce similar process maps could provide a useful basis for implementing information management strategies on construction projects irrespective of the method of procurement adopted.

Part C Improving the Strategic Use of IT in Construction

This final part of the book builds from the introduction of concepts in Part A and the assessment of current practice in Part B. It shows how the concepts, techniques and practical ideas we have introduced can be applied to move current practice further forward. This will be critical to construction businesses in the next decade as they seek to deal with the more competitive environment in which they work. The first two chapters in this part consist of simple tools and methodologies that companies can use in making the first steps forward.

These are followed by outlines of the short-, medium- and long-term research, development and innovation priorities for IT in the sector. This culminates in an overall vision of what IT might look like in construction in the first decade of the new millennium.

The part then concludes with two chapters that illustrate the importance of process management and organisation in allowing these technologies to become widely accepted in practice. The first

outlines a generic model of a radically redesigned generic design and construction process based on manufacturing principles indicating how various information technologies would fit within it. The final chapter of the book introduces the concept of capability maturity, illustrating how all of the ideas in the book will only make some impact if they are tied to an overall process of construction businesses becoming more systematic, learning, knowledge-based, process-oriented, virtual organisations.

As with all parts, the chapters in this final part are illustrated with managers' quotations and case studies and summarised with a one-minute manager's introduction.

13 A strategic health check

Martin Betts, Martin Jarrett and Mathew Shafaghi

> "IT can take up a lot of time and money if you need to give extensive training to staff... Many staff, particularly senior managers, still resist technology ... IT will only be effective if easy to use."
> – *Head of production preparation of Dutch contractor*

A health check for IT management embraces questions related to competition and business strategy, the role that IT is given in the organisation and the nature of the IT strategy. Responses to such a health check can help you determine the sort of company you are. Comparisons with what other companies do effectively provide an implementation plan for organisational improvement. Such a tool has been used by a number of construction companies, showing quite large differences.

- The key elements of an IT health check are: company and business strategy, the role of IT and the IT strategy.
- Four scenarios of overall IT management are distinguishable of companies that are: well behind their competitors, at a basic level of IT management sophistication, amongst the leaders in a number of areas, or close to strategic exploitation in many of the key elements of the process.
- Individuals within a company have different interpretations of how IT is managed.
- Poor current practice is reflected by the low seniority of involvement of IT in main board decision making and strategy generation.

> "We are very keen to make more effective use of IT. What we need to do first is better understand how to organise IT within our company."
> – *Chief executive, South African construction company*

13.1 Introduction

This final part of the book contains chapters that collectively form an insight into how leading organisations are embracing the future; the means of assessing how to get there; and an analysis of the IT innovation and research that may be necessary in moving more purposefully forward. Collectively, these are intended as a guideline for companies for the future. This first chapter is a health check of how IT is managed in companies towards strategic ends.

The findings from the studies of current industry practice in the previous chapters have been used in defining rules and guidelines which allow a company to identify how close to strategic IT exploitation it is. The first step towards more effective strategic exploitation of IT is to gauge the current position of your organisation by undertaking this strategic health check, based on the rules and guidelines presented here.

Each question that follows is accompanied by a grid drawn from rules and guidelines established from observing current industry practices, containing four possible scenarios that are indicated by letters D, C, B and A.

To carry out the health check, follow three steps.

(1) Select the scenario that reflects your company position from every question.
(2) Transfer your choice to the health check matrix in Section 13.3.
(3) Review your score.

Guidelines for improvement can be drawn from migration to higher scenarios in each scale. It should be borne in mind that this health check has produced general guidelines and recommendations relevant in certain circumstances. As all in industry will well know, the successful application of IT depends upon the right people being given the right opportunity at the right time in the right circumstances.

13.2 The health check questions

In Chapter 9, we described a process whereby a number of major construction organisations were studied in detail for the way they sought to strategically exploit IT. That chapter formed a description of current strategic practice. This chapter has a methodology for improving strategic practice. Analysis of data collected from industry resulted in over 200 propositions concerning the management of IT in construction organisations. Some of these propositions have been selected and translated into the following questions. We urge you to ask these questions of your business or businesses as a stimulus to your own debate about how you are using and managing IT. We hope it acts as a trigger to the ongoing debates about IT investment and management within your organisation. The questions are grouped into three categories relating to: competition and business strategy; the role of IT; and IT strategy.

Competition and business strategy

1 Does IT support your core competencies?

IT does not support competencies	IT offers some support to competencies	IT is critical to core competencies	Our core competencies are our IT systems
D	C	B	A

2 How can IT help your company to compete?

Changing our competitive behaviour	Changing competitive behaviour and supporting competencies	Changing competitive behaviour, supporting competencies and strategic alliances	Changing competitive behaviour, supporting competencies, strategic alliances and providing barriers to entry
D	C	B	A

3 What is the impact of IT on your corporate goals and objectives?

IT has no impact	IT has some positive impact	IT supports our corporate objectives through business efficiency and cost reduction	IT is critical to business efficiency, financial gains, engineering excellence, R&D and innovation
D	C	B	A

4 Considering IT, what do you perceive to be your position in relation to your competitors?

Twelve months behind other companies	Six months behind other companies	The same level as other companies	Ahead of other companies
D	C	B	A

5 What is the impact of IT on your clients?

IT does not have any impact	IT is important to some clients	IT is good for presentation and communication to clients	IT is essential for meeting clients' requirements
D	C	B	A

6 Do you think your IT expertise will help you to win work?

IT has no influence in winning work	IT is good for presentation	IT can be instrumental in winning work	IT is critical to winning work
D	C	B	A

7 How is IT used as a part of strategic alliances?

Spreading the costs and risks	Business efficiency	Improving competitive behaviour	To gain competitive advantage
D	C	B	A

The role of IT

8 How do you currently use IT in your company?

IT is not seen as an important part of the business	Use IT but let the technology find its own way within the organisation	Use IT after proven and satisfactory results from users and other companies	Proactively seek to use IT as part of a well-thought-through strategy for achieving competitive advantage
D	C	B	A

9 How would you describe the relation between IT and your business strategy?

They are completely separate entities	IT is indirectly addressed through its supportive role	IT plays a central role in our business strategy	IT is a critical success factor for national and global business strategies
D	C	B	A

10 What is the level of participation of your IT function in the formulation of your overall business strategy?

No participation	IT function simply informed of business strategy	Main board IT director responds to business strategy proposals	Main board IT director with part of the responsibility to suggest new business strategies
D	C	B	A

11 Does IT have an impact on your operational strategy?

IT supports operational strategy	IT helps to reduce costs through speed and efficiency	IT assists us to improve quality to meet clients' requirements within budget	IT enables us to provide products and services that are better than any one else's
D	C	B	A

12 Does IT have any influence on the delivery of your marketing strategy?

IT has no relation with marketing strategy	IT has some impact on marketing strategy	IT is important to the success of marketing strategy	IT is critical to the success of marketing strategy
D	C	B	A

IT strategy

13 How would you describe the use of IT systems in your company?

Mainly functional applications	Mainly functional applications with some integrated systems	Mainly integrated systems	Process support tools with full information sharing
D	C	B	A

14 What are the objectives of your IT strategy?

Business efficiency	Business efficiency, improving quality and improving relations with internal and external stakeholders	Providing support for business strategy	Delivery of business strategy
D	C	B	A

15 What is the thrust of your IT strategy?

Up-to-date IT facilities	IT skill level within the company and required IT facilities	Support for business strategy and user participation and empowerment	Exploiting IT for competitive advantages
D	C	B	A

16 How do you intend to manage IT in the future?

Follow other companies after assessing their experience	Develop IT capabilities at the same time as other companies	To experiment and develop the technology	To have advanced technology developed as a key part of the business
D	C	B	A

17 What are the critical success factors for sustainable competitive advantage through IT in your company?

We do not seek a sustainable competitive advantage through IT	Acquisition of right technology	To use IT for strategic and innovative opportunities that arise	Continuous R&D programmes and seeking innovative solutions to construction problems
D	C	B	A

18 How do you intend to manage IT projects and innovations?

We are not concerned with IT innovations	Out-sourcing	IT professionals and external facilitators	IT professionals appropriate management, user group committees, strategic alliances, internal and external facilitators
D	C	B	A

19 What is the level of R&D regarding IT in your company?

No interest	Planned R&D	R&D focus aimed at new future markets	R&D focus aimed at new future markets and the achievement of a sustainable competitive advantage
D	C	B	A

20 What is the nature of your IT department?

Small technical unit providing group technical service	Large technical group providing group technical service	Central technical support to individual business units	Fully developed IT responsibility for strategy and business responsibility
D	C	B	A

21 How much importance will you attach to IT skills within your company?

Low-cost training initiatives	Provide training when required	Seeking and upgrading the quality of training in our company	Develop intensive education and training programme for all IT users, including senior management
D	C	B	A

22 Who is aware of your IT strategy?

Technical staff	IT professionals and users	Business managers	All key internal and external stakeholders
D	C	B	A

23 What is the extent and nature of involvement of IT users in the formulation and implementation of your IT strategy?

There is no user involvement	Involved when required	Involved at the early stages	Involved throughout the whole process
D	C	B	A

24 What do you consider to be the risks associated with the implementation of your IT strategy?

Financial risks	Technological risks	Business environment risks	Strategic risks
D	C	B	A

25 How often do you review your IT strategy?

Every five years	Every three years	Every year	IT strategy is reviewed continuously
D	C	B	A

26 How do you measure your IT performance?

Short-term payback	Cost reducing aspects of IT	Value adding properties of IT	Strategic opportunities provided by IT
D	C	B	A

27 What are the characteristics of your IT strategy?

IT is dominated by financial issues	IT is driven from the bottom up	IT is driven by middle managers	IT is managed from the top
D	C	B	A

28 Who is the champion for IT projects in your company?

Technical staff	Functional managers	IT director	Senior management
D	C	B	A

13.3 The health check matrix

This allows you to transfer the choices from the strategic IT health check questions so that your profile can be recorded and distributed (see Figure 13.1).

13.4 The health check scenarios

These descriptions are the interpretations you might make of alternative scenarios that arise from completing the strategic IT health check.

(D) If you have mainly ticked scenarios in this category you are clearly utilising IT as a support tool for operational efficiency. It appears that IT is not viewed as an important part of your business and IT applications are mainly directed at support and functional systems with very little in terms of integration.

The financial justification for IT investments has prevented you from formulating a clear IT strategy in line with your business, market and industry requirements. You may need

Figure 13.1 Strategic health check matrix

Question	D	C	B	A
Competition and business strategy				
1				
2				
3				
4				
5				
6				
7				
The role of IT				
8				
9				
10				
11				
12				
IT strategy				
13				
14				
15				
16				
17				
18				
19				
20				
21				
22				
23				
24				
25				
26				
27				
28				
Total				

to analyse your current situation concerning various aspects of IT, including education and training programmes and to develop IT strategy that would focus on long term objectives and goals with consideration to your existing market status within your industry sector. The seniority and profile of staff involved with managing IT may be too low.

You need to rethink the way in which you intend to manage IT within your organisation. The supportive role of IT, the level of management commitment, low levels of IT skills and expertise, and the philosophy of following other companies concerning IT can no longer meet the needs of the dynamic and competitive environment of the late 1990s.

(C) If your company can mainly relate to the scenarios in column C of the health check matrix, you are aware of some of the strategic opportunities provided by IT, but your bottom-up approach has formed a barrier to your progress. Whilst it is encouraging to learn that IT is used to reduce costs and increase the overall business efficiency, you are not fully benefiting from the opportunities provided by IT. However, in this respect, you are better than some companies in your sector but behind many others. Therefore, you may need to analyse your situation compared to that of your competitors. This requires the involvement of the IT function in the formulation of your business strategy, more focus on training and better awareness of IT as a competitive weapon.

You may be beginning to plan to use IT strategically. But your perception of the level of risks involved together with cost and the complexity of technology is forming a blocker that is influencing the awareness of strategic possibilities offered by IT in your company. As you have already benefited from some of the opportunities provided by IT, you need to improve your industry position within your sector. This is only possible through being innovative, investing in people and the required technologies and finally by creating an IT infrastructure and culture that would support your competitive strategy and corporate goals.

(B) If you have mainly placed your company in scenarios in column B of the health check matrix, you are probably among the best-practice companies when considering the use and management of IT. However, there is still room for improvement. Whilst you are benefiting from some of the strategic opportunities provided by IT, the conditional commitment of senior management, limited involvement of

IT users, the low levels of IT skills in some departments and partial commitment to R&D initiatives may slow down the rate of progress of IT in your organisation.

You may be planning to strategically exploit IT for efficiency gains and competitive advantage. Whilst you may be ahead of some of your competitors, you are operating in a similar environment to them. Therefore to improve your market status and industry position you need to plan to use IT more aggressively to combat the threat of competition.

(A) If you can relate to the majority of scenarios in column A you are amongst the very few and best-practice companies that are truly exploring IT for its strategic opportunities. You are also in a position to benefit from the IT culture you have developed and maintain IT on the urgent agenda of concern of top management. However, to maintain and improve your business growth and the prosperity of your organisation, you should continue to do what you have been doing and focus on new and profit generating activities through R&D initiatives. You appear to be truly planning to exploit IT for its strategic opportunities. You need to maintain, guard and improve your position and to capitalise from your capabilities and expertise.

13.5 Summary

The health check has been used with a range of UK companies to assess how they measure up in regard to the four scenarios. In this summary section we give some overall assessments of how UK companies measure up, together with detailed descriptions of how four companies have used the health check as part of their strategic management and business improvement exercises.

The overall comments may be summarised as follows.

- There is a high level of current use of IT within companies in the context of strategies which are frequently reviewed and increasingly led by senior managers within companies.
- There is a low level of integration in current use of IT systems and little link between IT and business and marketing strategies.
- The greatest variability within the industry comes from the extent to which some companies are seeing IT as a strategic opportunity, with strategic rather than financial risks, and which requires active involvement with R&D.

The health check has been completed by 13 different companies and by more than 150 people. These include two major consultants, two client organisations and nine major contractors. From these responses, average scores for each of the 28 questions, and for the three major sections, were calculated. Standard deviations were calculated to show variability within the sample.

Those areas where UK industry comes out well overall are:

- current use of IT systems;
- intended management of IT in the future;
- intended management of IT innovations;
- frequency of review of strategy; and
- seniority of IT champion.

Those areas where industry comes out poorly overall are:

- participation of IT function in business strategy;
- influence of IT on marketing strategy;
- level of integration in current systems; and
- level of R&D in IT.

Areas of high variability between responses are:

- level of R&D in IT;
- user involvement;
- IT strategy risks; and
- IT strategy characteristics.

Four companies have used the health check in particularly innovative ways that are described here as a way of showing others what use could be made of this tool as a contribution to strategic management. The benefits of using the health check in these sorts of ways seem to be that it is a good:

- basis for workshops relating to use of IT by strategy groups within the business;
- trigger for initial and continuing discussion within the business about IT strategy;
- communications exercise in its own right; and
- way of identifying differences in perception between parts of the company.

Contractor 1

One major contractor completed the health check twice within three months. The first was as part of a workshop involving eight senior managers and directors immediately prior to a major strategy review within the company. This was mainly beneficial in initiating senior management debate to generate a collective answer to the questions. This debate was important in gaining general buy-in to the process of reviewing the strategy. The health check was completed a second time by two senior managers after a major strategy review. This ensured that the strategy proposed would result in major improvements to the way IT was to be managed by the company. Support for the strategy was taken from a major improvement in the company's health check profile from a second completion of the questions.

Contractor 2

This company had the health check completed separately by the group IT function, the civil engineering division and the special projects division. The very substantial differences in responses indicate how IT culture varies considerably between the business units of the major groups of construction companies. Reconciling this is a critical issue in more effective IT management within such companies.

Consultant 1

One major firm of design consultants had more than 100 senior managers complete the health check independently as a result of a mailing from the IT manager. Three different levels of seniority of senior management participated and their responses were separately calculated. They show significant differences in interpretation of where the company stands. These differences were reported back to a major meeting of all senior managers of the company. The exercise highlights major differences in understanding of the way IT is managed at different levels of seniority in the company and highlights communications problems and misunderstandings of responsibility for IT.

Consultant 2

This company had the health check completed simultaneously and independently through workshops involving members of senior management within different operating regions of the company. At the same time it was completed on behalf of the group management. Differences between the interpretations of the group and the separate operating regions and operating companies are being debated as part of the exercise of further development of a group-wide IT strategy within the company.

14 Measuring the benefits of IT innovation

Andrew Baldwin, Martin Betts, Derek Blundell, Karen Lee Hansen and Tony Thorpe

"Proposals for technology investments are always supported because the business never gives us the guidance to judge and assess business benefit."
– Group IT director, UK contractor

A tool for predicting and measuring the benefits of IT investment is a key mechanism for improving IT in construction. Tables can be used for making benefits predictions and post-implemention benefits measurements. These tables entail: selecting specific benefits for a particular IT project from a checklist of typical benefits, considering the 'do nothing' option, stating how each particular benefit is to be measured, allocating responsibility for achieving it to an individual, and defining each benefit in monetary, weighted score or qualitative terms. By using such a measurement tool for a new civil engineering estimating system, a company has found a better way of making the business case for IT innovation.

- IT benefits can be categorised into the three areas of efficiency, effectiveness and improved business performance.
- Measuring the benefits from IT requires: specifying the IT project, considering its strategic implications, understanding the business processes it affects, identifying and predicting benefits, implementing the solution, measuring the resulting benefits against the predictions and reconsidering the strategic implications.

"Our information systems shed some light on some problems but they do not resolve many of them. Management needs to know when things could go wrong and then take the right decision."
– Purchasing manager, US building contractor

14.1 Introduction

One of the biggest problems organisations have in making effective IT investments in construction is their inability to predict and measure the benefits that result. In many cases acts of faith are being followed. Some of the key questions that senior executives must ask themselves when sanctioning their acts of faith in new technology at present are:

- How can we justify spend on IT?
- Will it support our business strategy?
- Will it contribute value to our business processes?
- How will we know if we have achieved our aim?
- Can we measure a pay-back?
- Am I being conned by the boffins?

It is in the context of these key management questions that this chapter presents a practical methodology for predicting and measuring the benefits of IT investments. In considering an IT investment within an organisation, consideration has to be given to costs, technical issues, means of implementation, risk assessments, procurement strategy, and the likely benefits that will result. The last of these has typically been the most difficult to measure and communicate to senior business managers. This chapter outlines an approach to assessing ways in which IT innovation can benefit a construction business through improving:

- **efficiency** – doing things right;
- **effectiveness** – doing the right things; and
- **performance** – doing better things.

Throughout this chapter, these three concepts are made distinct. Efficiency is financially measurable and is represented by money. Effectiveness is measurable but not in monetary terms and is represented by improved operations. Finally, performance is not directly measurable in quantifiable terms but can be judged qualitatively by most senior managers who would recognise that IT can be of great consequence in influencing long-term business performance by increasing profit and market share.

The framework presented here is based on these three ways in which IT can improve your business. It is intended to be used by senior managers in construction to predict expected benefits of IT innovation and to gauge the benefits that result after their investment in innovation has been made. It should be used as part of the overall process of assessing new business investment proposals alongside currently used methods for assessing costs, judging risks and planning for training and implementation issues.

The term 'innovation' is used throughout this chapter and needs to be discussed. Innovation is often viewed as being something very special that is undertaken by very few within an organisation. We would argue against that interpretation. When implementing new IT, it is almost certain that some new ways of undertaking processes and activities will be necessary and different ways of conducting business will be caused. We would argue that any significant implementation of IT in a construction business should be treated like an innovation. Using the term 'innovation' might help focus readers of this chapter on the aspects of change and business improvement.

14.2 Using the framework

This chapter contains a number of sections that may be used to understand and apply the principles of measuring IT innovation benefit. These sections include:

- a diagram depicting the overall approach within the framework;
- general, strategic questions that should be asked within the business to understand the broad nature of anticipated benefits. These questions should be asked before and after an IT innovation and they have been completed with answers from a worked example;
- further details from a worked example of the use of the templates presented in Tables 14.1–14.3 for calculating the expected and actual benefits of an IT implementation;
- a means of making an overall assessment of the benefits to the business;
- an indication of additional issues that need to be considered when assessing and evaluating an IT investment; and
- a structured list of specific business benefits and improvements to business processes, which could be anticipated and measured.

A diagram depicting the sequence in which the parts of the framework are intended to be used is shown in Figure 14.1.

Figure 14.1 The measurement process

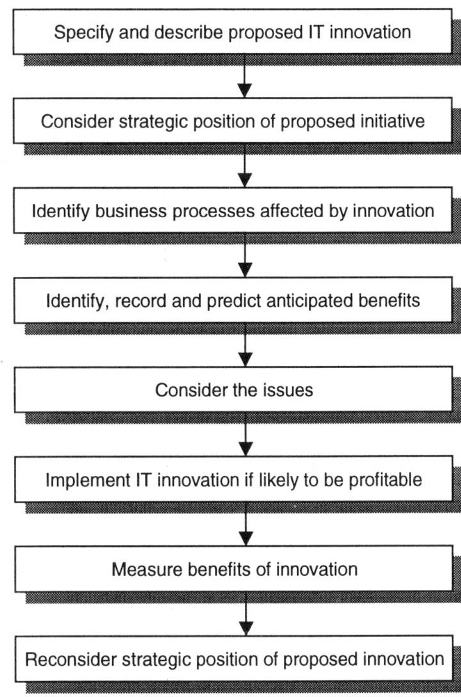

14.3 Business processes for a construction organisation

The approach that is taken within this framework is to suggest that innovating with IT can improve business processes within construction organisations. For the sake of this discussion, business processes have been defined as processes that cut through the entire organisation and that enable the organisation to exist as a business entity. Business processes referred to here are distinct from typical company functions or from parts or processes of projects. This choice has been made for two reasons: because many IT investments in construction enterprises are not related to specific projects but relate to infrastructure and head office processes; and, second, because the meaning given to names of functions differs considerably among contractors, consultants, subcontractors and suppliers. The framework presented is aimed at all these types of construction enterprises and the business processes chosen are generic. A definition of generic business processes of construction enterprises and a brief description of each follows.

Business planning: refers to strategic management of the organisation, deciding on new business ventures, and other senior management functions.

Marketing: refers to market selection, public relations activities, market intelligence, and generating new business.

Information management: refers to communication links between parts and locations of the business and external organisations, information archiving and distribution, and activities in information processing. It also includes information strategy and system planning.

Procurement: refers to all activities associated with the involvement of the organisation in the procurement activities of the client or customer and to the activities associated with procuring the services and activities of other participants to the organisation's input to the project. It therefore includes negotiation, bidding and tendering activities. It also includes procurement of products and services to support head office, non-project services within the organisation.

Finance: refers to activities associated with financial management and transaction processing on projects and at head office within the organisation.

Client management: refers to activities associated with managing relationships with customers, both on specific projects and in longer-term relationships.

Design: refers to those activities concerned with obtaining a brief, conducting feasibility studies, and sketch and detailed design activities undertaken on projects. It also includes non-project work in developing design technology and knowledge.

Construction: refers to those activities associated with production support on projects and extends into commissioning. It also includes non-project work in developing construction technology and knowledge.

Occupation and maintenance: refers to those activities associated with the use and occupancy, and maintenance phases of buildings and projects. It also includes non-project work in developing full life-cycle technology and knowledge.

Human resources: refers to activities within the organisation and on its projects that concern the management of people.

14.4 The worked example

In order to demonstrate how this framework may be used, a worked example is presented at the end of this chapter. The example presented has arisen from extensive testing and assessment of this framework with real companies. The results of the worked example have been documented and narrative comments added to aid interpretation by the reader. The sections in italics are explanation remarks added after the innovation was made.

The worked example is an unattributed, real example of the way the framework can be used and should be treated as an explanation of the approach within the framework. The results within the worked example are not intended to be a benchmark of comparable data for using this framework. There are no magic numbers that have to be reached that make IT suddenly profitable within a business. Rather, the framework provides a way of describing best practice in addressing the benefits prediction and subsequent measurement processes.

14.5 Applying the framework within a construction organisation

This work has been described as a framework rather than a methodology because each individual business will have to take the general principles of its design and tailor its use to their particular circumstances. Some readers may already apply such an approach from practices developed within their own business, or practices they have procured through management consultants. For them, this framework may act as a useful second opinion to ensure that as many factors as possible are being considered.

Others will have business analysts within their organisation who are well versed in such a general approach but may lack a specifically documented tool. For them, this framework may act as an appropriate *aide memoire.*

A third type of user may be completely unfamiliar with the processes and practices of considering IT investments, and may not have the available tools or experience base of people that are doing such work. For them, this framework may become more of a structured methodology to be applied by a senior business manager. We hope the framework has as many uses as possible but would stress that whatever the situation, the audience for the business case that is being made with the framework must be senior business management.

14.6 Specific IT benefits

The benefits which may accrue to a business through IT investment have been categorised as:

- efficiency benefits - doing things right
- effectiveness benefits - doing the right things
- performance benefits - doing better things.

The tables in the worked example provide an overall structure into which your specific anticipated benefits can be positioned, evaluated and assessed. Because of the nature of IT investments not all possible benefits may be identified initially and different benefits may emerge during implementation and use. This is part of the dynamic nature of a benefits prediction and measurement exercise and this framework has been developed to be sufficiently flexible to incorporate these instances as they occur.

14.7 Completing the tables

The left-hand column of Tables 14.1–14.3 identifies the business processes in which benefits may accrue. You should use your answers to Question 7 of the initial strategic questions to guide your selection, from the checklist at the end of this chapter, of those examples of typical benefits in each of the business processes that seem relevant to your proposed innovation. These are only illustrative of the types of benefits that may accrue. Having used the checklist as a guide you may then want to delete the column from the business case you are preparing with this framework as it might mislead the ultimate audience for the document.

The specific benefits arising from your proposed IT investment should be entered in Column 2 and the subsequent columns completed in accordance with the worked example. You will see that there are three tables corresponding to the three types of benefit that are possible from an IT innovation: efficiency, effectiveness and performance. The efficiency benefits require you to estimate financial values of benefits. The effectiveness table requires you to predict the likelihood and weight of the relative benefit of those that you have identified. The performance table requires you to rate and describe the impact of the performance benefit that you envisage. All three tables require you to consider what will happen if the innovation does not take place, to define how the benefit will be measured and to identify a person responsible for achieving and measuring the benefit.

The overall assessment of the business benefits includes a monetary sum from the expected efficiency benefits, a score from the effectiveness benefits, and a qualitative description and rating of the business performance benefits. These benefits are subjective to your organisation but can be used as indicators to enable comparison between competing investments or to show the relative impacts of a proposed investment. Over time, company benchmarks of scores can be compiled to better inform IT investment decisions.

14.8 Summary

This chapter contains a framework for predicting and measuring the different business benefits that arise from innovative application of IT within construction businesses. Senior managers' attitudes towards IT appear to treat it as a cost and as though it is a necessary evil. There is not a widespread perception that IT has delivered business benefits to organisations. We hope this framework will convince senior managers within construction that there are more potential benefits than might, at first sight, appear to be the case. We also hope it will help organisations to become specific about measuring the benefits that they are achieving with their use of IT.

A worked example

Description of innovation

Describe the background to the IT innovation that is being considered, some of the technical issues involved and, most importantly, identify the people and parts of the business concerned with the proposed innovation.

This proposal for IT innovation within the business is for the rewriting of the estimating system being used within the civil engineering business of a major international construction firm. This is being proposed to make the internally developed software suitable for a Windows environment.

The incentive for making this change is coming from the estimating department and from users of the system within the company.

Some of the other important issues concerned with the innovation include making a new system CITE compliant, making software suitable for a standard PC platform, and increasing data integration between different parts of the business.

Some of the different parts of the business affected by the IT innovation are estimating, purchasing, contracts management, procurement and group IT.

The champion for the project is the managing director of civil engineering. As champion, the managing director takes overall responsibility for the innovation and ensures it has adequate senior management support.[1]

The person responsible for ensuring the innovation delivers business benefit overall is the chief estimator within civil engineering.[2]

The system was successfully implemented over a slightly longer timescale than had been anticipated. The costs of the implementation were greater than those that were envisaged at the start of the project. The benefits that were measured overall were equal to or greater than those anticipated at the start.[3]

Notes

(1) We urge all companies to identify a senior manager within the business to take corporate ownership of each major IT innovation to ensure it has a business benefits focus and to allow other business and process changes to occur to allow these benefits to be realised.

(2) We urge all companies to identify a manager directly involved with an IT innovation who more specifically takes executive responsibility for the implementation achieving its expected business benefits.

(3) It is vital that the implementation is reviewed against the original expectations at the level of the description of what took place, the strategic issues it addressed and the specific benefits that it realised. This should be planned for as there will be a long time lag between it being filled in for expectations and for measurements, and there are likely to be personnel and planning changes in the interim.

Strategic questions

(1) What is the business opportunity/need that has triggered the idea of an IT investment?

> The business opportunity/need is to increase flexibility in the estimating department to allow more responsive bidding and greater data integration within the process of winning work. The specific triggers are:
>
> (1) The current estimating system runs in an obsolete Xenix environment and data interchange is difficult and only possible via specially written export routines.
> (2) The current application does not meet CITE standards for estimating systems and cannot accept electronic input.
>
> *(1) The system is now fully running in a Windows environment within networks operating within the company. Internal information interchange has been made possible by a combination of e-mail attachments and use of internal servers.*
> *(2) The new system fully meets the CITE standards for estimating systems and some level of electronic exchange now takes place during estimating with suppliers and subcontractors.*

(2) What strategic aim or business objective does the IT investment meet?

> The strategic aim is to have a widely accessible estimating system for the business that allows internal and external electronic links. More specifically:
>
> (1) The new application will run on a 'standard' PC configuration, will make data easily accessible and allow hardware and software to be easily maintained.
> (2) The new application will comply with CITE.
>
> *The new application has been widely installed on PCs throughout the business and has allowed integration.*

(3) How does the proposal for the IT investment fit within the current business plans for different parts of your organisation?

The current business plans for the group are focused on increasing profitability by expanding contracting work within core areas of project activity and by reducing overheads throughout the group. The implications for civil engineering are to develop the business to contribute towards both increased turnover and reduced overheads.

The place of this particular IT innovation in regard to these business plans is:

(1) It will help minimise overhead costs by reducing maintenance cost of software and hardware.
(2) It will reduce staff costs when CITE implemented.
(3) It will minimise retraining costs as system functionality and process is as per current system.

These are all elements of the current civil engineering business plan which seeks to broadly expand and develop new business, rationalise overheads, make links with industry-wide initiatives and exploit the current and future abilities of our people.

Subsequent to implementation, the issue of information links with clients has become more important as revisions to the business plan have made partnering and client links a more important issue.[1]

Note

(1) In reviewing the innovation against the business plan, the way the latter will inevitably have changed over the period of the implementation should be considered.

(4) How does the proposal fit within the current IT strategies for different parts of your organisation?

Part of the current IT strategy within the civil engineering business is to increase the level of penetration of a standard configuration of PC technology to all users within the business. Other aspects of the current IT strategy include: standardisation of software around the Microsoft Office Automation suite and the Windows interface, and adoption wherever possible of industry standards for data exchange and electronic trading.

With regard to this, specific aspects of this proposed IT innovation which relate to the current IT strategy are:

(1) The hardware and software configuration will comply with company 'standards' for ease of support.
(2) The new estimating system will be CITE standard supported.

Group IT strategy is not related to or affected by this proposal.

Subsequent to the implementation, group and business unit IT strategies have become more closely concerned with electronic links with clients.

(5) How does the proposal meet the needs of clients and/or strategic business partners?

Strategic business partners are key to this area of activity as estimating involves extensive consultation with suppliers and subcontractors and data input from the client and consultants.

With regard to downstream partners, the positive aspects of this proposal are:

(1) It is CITE enabled and will therefore allow further development of electronic links with key suppliers and subcontractors.
(2) If we regard sites as customers of estimating, then estimating can offer a better service as data are more readily available in an electronic format. Benefits can then be passed through to the sites.

With regard to upstream partners the further development of the estimating system will allow it to be amenable to direct input from upstream design, specification and bill of quantities information should electronic transmission of such data become the norm.

The importance of links with clients and strategic business partners has become more central to the proposal during the period of implementation.

(6) What would be the implications to the business if the proposed IT innovation was not made and nothing was changed?

The current estimating application cannot be sustained indefinitely in its current environment. As estimating is a core business system, a replacement strategy is required as this inevitable demise of the application will seriously harm the ability of the company to complete tenders if no action is taken.

This would have major implications to the business as a whole in terms of our continuing efforts to secure work in an increasingly competitive environment.

(7) Overall, how does the proposed IT investment affect the following business processes?

Business planning
Implications of the investment to business planning were identified after the innovation was made.[1]

Marketing
Ability to produce high quality reports.

Information management
Ability to exchange information with business partners.

Procurement
Ability to produce competitive tenders to win work.

Finance
Ability to provide data to the cost and recovery monitoring programme.

Client management
Ability to produce competitive tenders.

Design
No implications to design process were anticipated.
Sharing of data from one project to the next was found to be possible following the innovation.

Construction
Infrastructure to support improved information exchange with internal and external construction team.

Human resources
Some limited training implications.

Note

(1) It is important to be aware that parts of the business process affected by an IT innovation may turn out to be different from those expected.

Table 14.1 Business efficiency benefits

					Expected benefits			Measured benefits	
1. Business process	2. Specific benefits	3. Implication to this benefit of not making the innovation	4. Means by which benefit will be measured	5. Person responsible for achieving and measuring this benefit	6. Monetary value (£)	7. Likelihood of benefit occurring (%)	8. Expected benefit, Col. 6 × Col. 7 (£)	9. Specific benefit resulting	10. Monetary value (£)
Business planning									
Marketing									
Information management	No software licence fees[1] Reduced maintenance	Purchase new software Maintain 386 PCs	Cost of licences, less software rewrite Saving on maintenance contracts	Civil engineering IT manager[2]	15 000 4000	100 100	15 000 4000	*All benefits realised as anticipated*	*15 000*[3] *4000*
Procurement									
Finance									
Client management									
Design									
Construction									
Operation and maintenance									
Human resources	Reduced training	More staff training	3 days training/staff @ £100/day	Head of estimating department	12 000	100	12 000	*Benefits realised as anticipated*	*12 000*
			Total				£31 000		*£31 000*

Notes
(1) The specific benefits identified will relate to some of the items in the typical benefits checklist but not all.
(2) Achieving benefits is the overall responsibility of the champion for the innovation but a number of individuals within the business will then share different parts of the responsibility for ensuring that they are achieved.
(3) Financial benefits in efficiency should be measured after the implementation in monetary terms. They will not always be exactly as those expected.

Table 14.2 Business effectiveness benefits

					Expected benefits			Measured benefits	
1. Business process	2. Specific benefits	3. Implication to this benefit of not making the innovation	4. Means by which benefit will be measured	5. Person responsible for achieving and measuring this benefit	6. Likelihood of benefit occurring (%)	7. Weighting (column total = 100)	8. Predicted benefit (Col. 6/100) × Col. 7 (max. = 100)	9. Specific benefit resulting	10. Measured benefit (max. = 100)
Business planning	Ability to cut and paste data from Excel and Word. This will increase the productivity of estimators and will raise the number of estimates that can be handled by the department. Depending upon workload, this may result in fewer staff being required	Cannot cut and paste between Windows and estimating package	Number of estimators copying data from side sheets directly into makeups	Estimating office manager	50[1]	20[2]	10[3]	*Increased productivity of estimators has occurred which has been of greater weighting than anticipated.*[4]	25[5]
Marketing	Submission quality reports, including company logo. This will remove the need for reports having to be re-typed for transmission to clients	Existing reports of poor quality – not fit to submit to client	Submission of priced bill to client direct from estimating system	Chief estimator	10	10	1	*No measurable benefit resulted*	*0*[6]
Information management	All reports now on A4 laser paper. This will remove the need for reports to be retyped for transmission to clients	Only output to dot matrix printers Reports do not fit in lever arch files	Scrapping of dot matrix printers Reduced filing space	Civil engineering IT manager	100	10	10	*Improved administrative procedures*	*10*
	Ability to submit same number of tenders as previously. The continued operations of the estimating department without having to employ additional people will be possible	System failure could prevent submission	Logging of current system failures	Estimating office manager	10	10	1	*Number of system failures has halved. This is of greater benefit than expected*	*15*

Procurement	Ability to analyse tender to best advantage. This will improve the information available at tender stage that may benefit the business by increasing the success rate for bids and ensuring that those tenders won are more profitable jobs for the business	Obsolete hardware will detrimentally affect our ability in future	Production of new analytical data at bid stage in future	Chief estimator	100	10	10	*Improvements in ability to analyse tenders has been achieved but the relative benefit is less than expected*	*5*[7]
Finance									
Client management									
Design									
Construction	Ability to output data in table format for site use in Excel. This will improve information made available on sites, improving site cost control. It will reduce the amount of re-keying of data at this stage	Cannot produce data in Excel format, must go through CSV	Number of new routines which are created for increased data export	Contracts manager	10	40	4	*Some extra output facility has been realised but of less benefit than expected*	*20*
Operation and maintenance									
Human resources									
			Total			100	36		*75*

Notes

(1) It is important to make some prediction of the likelihood of all types of benefits occurring however approximate this may be. Feedback from previous applications of this sort of framework will be useful for future uses but judgement and 'hunches'are likely to be important contributors.

(2) For effectiveness, all specific benefits should be weighted before and after the implementation so that relative assessments of impact to the business can be made. The weightings you apply should be your assessment of the relative importance of each of the effectiveness benefits. You should apply total weightings of 100 whether your list of effectiveness benefits contains 1 or 21 items.

(3) The predicted benefit can be assessed by multiplying the likelihood and weighting. This will generate a number whose use is a benchmark against which 'after the event' comparisons can be made. These numbers do not mean anything in themselves and there is not a target figure that can be aimed at. Users might find these numbers useful comparisons between alternative investments. Within a single investment, these individual numbers might be useful in helping to understand the spread and focus of value that is expected by the business in managing implementation.

(4) Some qualitative assessment of the nature of the efficiency benefit compared with that anticipated should be stated.

(5) After implementation, relative weightings of efficiency benefits may turn out to be different from those expected, shown in Column 8. If some specific benefits have become relatively more important, others will have to be judged less important. The overall column total for Columns 9 and 10 must be 100 at the most. The total for Column 8 must be 100.

(6) If a benefit was expected but not realised this must be indicated and recorded.

(7) This is an example of where a benefit was realised but its weighting is judged as less important than had been expected.

Table 14.3 Business performance benefits

					Expected benefits			Measured benefits	
1. Business process	2. Typical benefits	3. Specific benefits	4. Implication to this benefit of not making the innovation	5. Means by which benefit will be measured	6. Person responsible for achieving and measuring this benefit	7. Likelihood of benefit occurring (%)	8. Qualitative rating and description of the impact of the expected benefit A = Very significant B = Significant C = Moderate D = Low	9. Specific benefit resulting	10. Qualitative rating and description of the impact of the measured benefit A = Very significant B = Significant C = Moderate D = Low
Business planning	Providing space and capacity for business growth Safeguarding future flexibility Overcoming obsolescence Increasing responsiveness of senior management to business problems							*A benefit regarding future flexibility has been realised that was not anticipated*[1]	*A. The company has found that its new estimating system has increased its ability to work with new and repeat clients by allowing direct electronic links for design information*
Marketing	Improved strategic intelligence for new markets Improved PR targeting and delivery								
Information management	Improved information version control Ease of capture of meaningful information More relevant and reliable data Improved filtering of information	All data held in Windows format Greater integration with other systems More accurate data entry Systems on common platform Scrap obsolete platforms		No. of new reports No. of systems which link to estimating Electronic input Links to estimating No. of systems scrapped	Chief estimator	10 50 50 100 100	B. These benefits will bring overall improvements in the way information is managed within the estimating process. The implications will be greater flexibility in estimate production, increased accuracy and reduction in diversity of systems. These benefits are likely to be important to business performance improvements as they directly relate to current business plans[2]	*The links to estimating and reduction in new systems have been realised. There has not been an increase in the no. of new reports but a much improved facility for electronic input from other systems which link to estimating*	*B. The measured benefits of increased flexibility, accuracy and reduced diversity have all been realised and are substantially improving business performance*

Procurement	Improving external access to stock levels and price information More effective identification and assessment of new suppliers	Can receive and transmit BoQ in CITE format		Receipt of electronic bill	Chief estimator	10	D. There is a small chance that the system will allow direct electronic receipt of bid documents from clients. The obstacle to this happening is that few clients currently operate in this way. The implications to business performance if they did would be substantial	*Following implementation, a large number of clients have now started to issue bid documents electronically*	*A. This has dramatically improved the company's business performance and the installation of the estimating system has retrospectively been seen to dramatically improve the company's strategic position*[3]
Finance	Improved/new transaction methods Improved forecasting and control Greater integration with other functions								
Client management	Improved information exchange with clients Increased client satisfaction Strategic competitive advantage								
Design	Improved idea sharing among project team Improved integration							*The ability to input data from previous projects has arisen*	*B. This has become more important as a benefit to the company with the way that business and IT strategies have changed since implementation*
Construction	Improved idea sharing among project teams Improved integration Improved project relationships with strategic partners	Sites will be able to receive sets of defined data electronically		Number of new data sets defined	Contracts manager	50	C. Some capture of real cost data from sites is anticipated as a means of monitoring the accuracy of estimates	*None*	*D. No improvement in this process has occurred*[4]

Contd

Table 14.3 *Contd*

					Expected benefits			Measured benefits	
1. Business process	2. Typical benefits	3. Specific benefits	4. Implication to this benefit of not making the innovation	5. Means by which benefit will be measured	6. Person responsible for achieving and measuring this benefit	7. Likelihood of benefit occurring (%)	8. Qualitative rating and description of the impact of the expected benefit A = Very significant B = Significant C = Moderate D = Low	9. Specific benefit resulting	10. Qualitative rating and description of the impact of the measured benefit A = Very significant B = Significant C = Moderate D = Low
Operation and maintenance	Improved capture of design and construction decisions Improved full life-cycle information management								
Human resources	More effective assembly of project teams Enabling of cross-functional teams Improved human relations Regularised working arrangements								

Notes

(1) Performance benefits are intangible and highly subjective. It is very likely that some benefits will be realised that were completely unforeseen when the innovation was planned. These tend to get taken for granted. We advise you to think very carefully about instances like this at the end of an IT implementation. They are an important way in which IT benefits tend to get undervalued.

(2) Different organisations using this framework will decide on these ratings in their own way. No standard definitions of these ratings are provided as it is what these words mean to your business that is important. It is advisable that you aim to apply the same interpretation of these words every time you use this framework.

(3) The rating of an intangible performance benefit could be very different from what is expected. We suggest you use the four-point rating scale to make this distinction clear and try to record some words that help clarify to people within the business the nature of the performance improvement achieved.

(4) There will of course be other instances where the performance benefit is less than expected. Again the rating scale and descriptive words should be used to help make fair judgements within the business.

Table 14.4 Overall business benefits

Types of benefits	Expected benefits	*Measured benefits*
Efficiency benefits – quantifiable and valuable	Total forecast monetary value £31 000	*Total realised monetary value £31 000*[1]
Effectiveness benefits – quantifiable but non-valuable	Total forecast score/100 36	*Total realised score/100 75*[2]
Business performance benefits – non-quantifiable and non-valuable	C. The performance benefits expected are in the area of improved information management, increased flexibility in electronic integration and links between site cost data and estimating	*B. Performance benefits that were realised were in improved flexibility of information management and much improved electronic links with clients. The expected performance benefits relating to site links were not realised*[3]

Notes

(1) This figure is the easiest to calculate but should never be used in isolation as an assessment of overall benefit.

(2) This number is purely indicative and relative. There is no threshold value that should be used and comparisons of this number between different projects should be made with great care and sensitivity. The use of a number here is only indicative in making comparisons of effectiveness benefits before and after an IT implementation.

(3) This overall rating is based upon taking a rough overview of the ratings of individual specific benefits in Table 14.3. The description in Table 14.4 compared with that in the expected benefits column is useful in making an overall assessment of how measured performance benefits compare with those that were expected.

14.9 The experience of the worked example

Some features that the company behind the worked example have seen in their use of the framework are as follows.

- Traditional cost benefits fit easily under the heading of 'Efficiency'.

- Checklists are a great help for the softer benefits.
- The means by which benefits are measured sets clear accountability.
- Predicting likelihood and weighting discourages over-expectation.
- Performance benefits really force strategic thinking.
- Previously unquantified benefits are identified by using this process.

The results of using the framework on this worked example to the company concerned were as follows.

- New benefits resulted.
- A focus was given on value and tomorrow's opportunities.
- It created a learning culture and a framework to build on.
- It triggered new thinking and process change.
- It demonstrated a professional and consistent method.
- It discouraged subjectivity.
- It provided a means for comparing alternative proposals and solutions.

14.10 Checklist of typical benefits

Business process	Typical efficiency benefits	Typical effectiveness benefits	Typical performance benefit
Business planning	Reduced planning times	Increased sales Minimising business risk Strategic competitive advantage Increased business flexibility Maintaining competitive capacity Reduced risk in new business ventures	Providing space and capacity for business growth Safeguarding future flexibility Overcoming obsolescence Increasing responsiveness of senior management to business problems
Marketing	Reduced marketing costs Ability to handle more enquiries	Improved company image Generating new business Increased market share	Improved strategic intelligence for new markets Improved PR targeting and delivery

Contd

Business process	Typical efficiency benefits	Typical effectiveness benefits	Typical performance benefit
Information management	Reduced communications costs Reduced paperwork Reduced IT costs	Easier international links Fewer information bottlenecks Improved quality of output Sustaining market share	Improved information version control Ease of capture of meaningful information More relevant and reliable data Improved filtering of information
Procurement	Reduced storage requirements Reduced transaction times Reduced transaction costs Improved delivery scheduling	Maintaining competitive capacity Faster response to supplier quotations Ability to provide instant price quotations to clients	Improving external access to stock levels and price information More effective identification and assessment of new suppliers
Finance	Faster invoicing Reduced transaction costs	Minimising business risk Better control of cash flow Reduced lead times for financial reporting	Improved/new transaction methods Improved forecasting and control Greater integration with other functions
Client management	Quicker response to client enquiries Quicker response on current project progress	Improved quality of output Faster delivery of services Improved focus on client requirements	Improved information exchange with clients Increased client satisfaction Strategic competitive advantage

Contd

Business process	Typical efficiency benefits	Typical effectiveness benefits	Typical performance benefit
Design	Reduced lead times for design Reduced rework Increased information exchange	Improved quality of output Reduced technology risks More responsive ability to arrange meetings Increased speed of new design development	Improved idea sharing among project teams Improved integration
Construction	Reduced construction times Improved productivity Reduced waste	Improved quality of output Reduced technology risks Ability to exchange data	Improved idea sharing among project teams Improved integration Improved project relationships with strategic partners
Operation and maintenance	Reduced operating costs Quicker access to operation and maintenance data	Improved quality of output Ability to refer back to data	Improved capture of design and construction decisions Improved full life-cycle information management
Human resources	Reduced staff requirement Reduced training requirements	Improved record of staff skills Improved ability to select appropriate team members	More effective assembly of project teams Enabling of cross-functional teams Improved human relations Regularised working arrangements

15 Priorities for IT innovation

Martin Betts, Andy Clark, Antonio Grilo and Marcela Miozzo

"We have to drive the waste out of construction."
– Procurement director, global construction client

One way of identifying how IT can add value to a construction business is to follow a process analysis approach. By analysing the main blockers in the generic process, and seeing what needs to be done to overcome them, an agenda for IT innovation emerges. Doing this for key parts of the project process generates general ideas for IT-enabled process improvement and many specific suggestions.

- Innovation priorities for construction to overcome process blockers includes use of visualisation and intelligence technologies to address buildability issues, and communications technologies to improve supply chain management.
- Design innovation priorities include use of intelligent systems for better design reuse and design intent capture, communication technologies for improved team working, and integration technologies to allow concurrent design processes.
- Briefing innovation priorities include: intelligence tools to improve business process analysis, and communication and integration tools to allow concurrent briefing processes.
- Facilities management innovation priorities embrace the use of visualisation and intelligence technologies to allow industry data sharing and inspection and space planning of facilities.

"I am sure there is so much more we can do with IT. Our biggest problem is knowing where to start."
– Chief executive, Australian construction contractor

15.1 Introduction

This chapter sets out to show the innovations that companies in the industry should make to improve the process performance of the industry for its clients. It does this by defining a generic process structure for the commercial construction sector. Processes are defined as a set of value-adding linked activities, cutting across specialisations, following the flow of material and information, from raw materials to end products. A detailed study of these different processes enables the identification of inhibitors against smooth process flow (hereafter called *blockers*). These are characteristics or activities that fail to add value to processes, contribute to defects, delays and waste, and operate against designing quality into processes and the coordination of different activities. An attempt is made to identify the source of these obstacles and to provide suggestions for their elimination (hereafter called *eliminators*). This is accompanied by a draft of a workplan aiming to create a coherent industry-wide innovation strategy for construction with the place of IT research and innovation separately identified.

15.2 The generic project process

The following analysis is largely based on case-study material from interviews with managers of British Airports Authority (BAA) and a workshop by participants with current industrial and consulting experience in the architecture, building surveying, quantity surveying and construction management areas. BAA, an important client to the construction industry in the UK, has embarked upon a programme of construction process change. This includes the initial establishment of 12 out of 200 planned framework agreements with suppliers and partnership agreements with contractors and designers; a move to design standardisation and computer prototyping of buildings before the start of construction works (e.g. North Terminal Domestic lounge for British Airways at Gatwick, UK); and palletised, bar-coded delivery of construction components. These initiatives are part of an effort to substitute a highly integrated production line process in construction (feasibility, concept, prototyping, manufacturing, logistics and site assembly) for the traditional system of producing buildings (feasibility, concept, design, tender, construction) (NCE/NB, 1995). BAA have also documented their interpretation of a generic process protocol in the form of a handbook for their project services staff, a wall chart of the process and an explanatory video-cassette.

In this chapter, we argue against the project stages adopted by the Royal Institute of British Architects (RIBA), in their long established Plan of Work, as the basis for a generic process structure of commercial construction. This defines inputs, outputs and objectives solely as a basis for fee payment. The generic process structure for the commercial construction industry used in this chapter is an amended version of BAA's attempt to identify their different construction processes to allow more general applicability to other firms and organisations in the sector. An initial taxonomy of generic core and support processes is developed according to their importance for construction in adding value and converting material from one state to another. Core processes, in order of importance, include: construction and manufacture, design, project definition, business analysis and facilities management. Support processes include: evaluation, project execution, procurement and cost management. Core processes make an important contribution to the production process in building, while some of the support processes are a combination of productive activities and transactions (evaluation, project execution) and others involve mainly transactions (procurement, cost management). Each of these core and support processes consists of sub-processes (see Table 15.1).

15.3 Process improvement potential

Each of the core and support processes and their sub-processes was studied in detail to identify obstacles, sources of these problems and ways of eliminating them and to generate a workplan for construction industry change as shown in Tables 15.2 to 15.9. The proposed workplans are highlighted in the last columns. For example, Table 15.2 illustrates the main obstacles to the flow of the construction and manufacture process. The problem of poor supply chain management, arising in particular from the processes of materials procurement and industry fragmentation, can be eliminated by studying best practice in supply chain management inside and outside construction and by an initiative by the industry actors to implement IT supply chain coordination practices. Many new industry and research initiatives in areas such as EDI, as evidenced by EDICON and CITE, and bar coding are attempting to deliver this best practice.

Table 15.1 Construction processes and their sub-processes

Processes	Sub-processes
Core processes	
Construction and manufacture	Resource procurement, facilities mobilisation, construction monitoring, construction execution, materials management, subcontractor logistics, finalisation and commissioning
Design	Design analysis, design synthesis, design appraisal, design decision, design review, production of design deliverable, design standards, experience management, design for manufacture
Project definition	Needs identification, briefing, scope definition, formation of consultant team
Business analysis and facilities management	Customer assessment, market research, business requirements analysis, development of business case
Support processes	
Evaluation	Risk assessment, option evaluation, 3D prototyping, testing and feasibility, performance evaluation, feedback
Project execution	Project evaluation planning, programming, progress measurement, risk management, health and safety, schedule management
Procurement	Strategy options, procurement of standard components, use of preferred suppliers, contract administration, drafting of contract
Cost management	Cost modelling, cost planning, cost estimating, cost reporting, forecasting, contingency management, life cycle cost management, value management

Table 15.3 illustrates the main obstacles that hinder the design process. The problem of requiring staged design presentation due to the traditional design fee system could be eliminated by an industry initiative to define payment and process performance on the basis of overall value contribution rather than discrete sub-process optimisation. This requires an industry effort to redefine the project plan of work and the consultant fee stage payment guidelines. Professional deregulation and many other major organisational changes affecting the industry, in particular the rapid emergence of design and build, multi-disciplinary design, and project management, are attempts to challenge the traditional plan of work. In a similar way, the emergence of the CIC (Construction Industry Council) as a multi-professional body in the UK and the emphasis of the UK Latham Report (1994) and Egan Task Force Report (1998), on team-building, lend particular support to this form of process change.

Table 15.7 similarly illustrates the problems in the project execution process. The obstacles to project coordination arise from the contractual boundaries and divided responsibility for specialist activities. Efforts to allocate 'virtual organisation' responsibilities and to share information in a team would eliminate these problems. However, such efforts are contingent upon an industry initiative which would remove contractual and intellectual barriers through establishing contracts and teamwork in order to set up the virtual organisation structure which follows best practice, both inside and outside the industry. Again, a major part of the UK Latham Report recommendations and its focus on implementation of contractual changes reflects this concern as does Egan's calls for integrated design and construction processes.

Much of the support to the changes resulting from this process analysis in construction management thinking and the emerging research agenda has been documented. For example, the emphasis on the need for improved project briefing and partnering reflect issues raised in the influential UK Building 2001 studies (Centre for Strategic Studies in Construction, 1988). The emphasis on project information integration, knowledge-based engineering systems, and the adoption of virtual reality as a key technology also support major recommendations that emerged from the Technology Foresight in Construction report implemented by the UK Office of Science and Technology (OST, 1995). Other recommendations echo observations resulting from strategic analyses of professional structures such as the redefinition of professional roles (Centre for Strategic Studies in Construction,

1988; Betts and Ofori, 1994). The fact that these, and other process change priorities, are generated by the type of process analysis being conducted in this study gives credibility to such an approach and its ability to generate a coherent process innovation strategy for the construction industry.

Tables 15.2 to 15.9 outline this approach and illustrate its general usefulness in identifying problems which arise in one process due to practices in other processes. The tables also represent a draft workplan for a whole industry initiative of process change. In particular they identify likely priorities for construction IT research and innovation that would most directly address industry improvement needs. They may be directly usable as priority areas for innovation by construction companies seeking to strategically exploit IT.

15.4 An overall process improvement work plan

The outcome of this process analysis is the definition of a framework and a workplan for an IT-enabled process strategy for the UK construction industry to which individual companies might respond. This framework includes:

- an account of critical success factors for the industry;
- the key business issues faced by the industry;
- the identification of core and support processes in construction;
- the nature of the process obstacles in construction;
- the nature of construction-specific solutions to these obstacles;
- separate identification of IT-based solutions; and
- a comparison with current IT applications.

The workplan resulting from this study is a comprehensive picture of innovation activities that are required to facilitate the flow of information and material in the construction processes. The specific priorities for IT research within construction are also identified. This workplan has elements addressing:

- the transfer of process-based methods applied in other industries to construction;
- the use of multi-disciplinary and cross-industry workshops to identify economic and technological problems and opportunities in construction;
- the implementation of benchmarking activities and identifica-

tion and dissemination of best practice in the industry and in other industries;
- activities leading to the adoption of standards across the different organisations in the industry;
- the implementation of pilot developments and their follow-up;
- the introduction of organisational changes; and
- the commissioning of research projects.

The details of this workplan can be found in a more comprehensive document (Construct IT Centre of Excellence, 1995b).

At this point it is relevant to consider who would find the outcomes of a study like this actionable. In earlier chapters we have argued that IT applications in construction can be made at a number of different levels. These include the basic operating levels of the sector that embrace its products and the major projects undertaken. It would also include corporate applications more typically considered in other sectors. A level of application relevant to this work is the national level applications and coordination of IT. National strategies for IT are becoming increasingly common as we have seen and have been widely reported (Mathur, Betts and Tham, 1993). This work forms part of an ongoing initiative within the UK to coordinate and direct IT research and innovation towards a common process-based research agenda.

15.5 Summary

This chapter has attempted to apply a process analysis to the construction sector. Underpinning the analysis is a dynamic characterisation of a process strategy, which draws on some of the concepts of process-based methods used in manufacturing and current process-based initiatives in construction practice. Application of these concepts enables the definition of a generic process structure for commercial construction that identifies blockers and eliminators and outlines a workplan for industry change. Contrary to the argument that the new management methods found in manufacturing are inapplicable to construction, this study has produced construction-specific techniques and solutions to the problems and peculiarities of the industry.

The result has been the definition of a draft framework of an IT-enabled process strategy and a workplan for a coherent strategy for process change in the construction industry. This framework includes the identification of obstacles and the suggestion

Table 15.2 Construction and manufacture process

Blocker	Blocking source	Eliminator	Workplan
Constructability	Design	Standardisation Prototyping	Process team design **VR and CAD prototypes that allow production evaluation** Design standards **Knowledge-based systems (KBS) and case-based reasoning (CBR) of building solutions**
Design changes	Design	Design freeze Deliverables	Project stage deliverables Design standards Design risk management systems
Uncertainty of tender outcomes	Procurement	Alternative procurement route Partnering	Process-oriented procurement systems Partnering and virtual organisations
Coordination of main and subcontractors	Project execution Procurement	Effective planning Lean production JIT	Implementation of lean production Implementation of JIT in construction
Poor supply chain management	Industry fragmentation Materials procurement	Effective supply chain logistics JIT	**IT supply chain coordination prototypes** Best practice in supply chain management from inside and outside of construction Pilot studies of JIT systems in construction Procurement support

Sequential working	Procurement	Lean production KBS planning	Lean production in construction Systems of concurrent engineering in construction Benchmarking of good production management practice **KBS planning systems**
Temporary project structures	Fragmented industry Lack of defined project structure	Partnering – approved suppliers Virtual organisation structure	Process and partnering procurement benchmarking Research into virtual organisation structures Accreditation and approval databases
Late appointments	Procurement route	New procurement structures Partnering relationships	Project stage deliverables Virtual organisation structures

Items in bold are specific IT research topics.

Current use of IT in the construction and manufacture process

This chapter both presents a view of future IT research need, and relates it to a summary of current IT practice. This review of practice is substantially facilitated by the major best practice IT benchmarks summarised in Chapter 11. The subjects of those studies do not map directly to the process structure presented here. The former is process as it currently stands. The latter is process as modern management theories suggest it should be.

The benchmarking exercise that is primarily relevant here is that relating to construction site processes (Construct IT, 1996a). This shows typical overall construction practice to be a long way from world class practice in engineering. The exception is IT use to support planning, monitoring and control where construction practice is world class, emphasising the strong project management culture within the industry. The areas where construction is furthest behind other sectors are: handling materials and deliveries; legal and health and safety issues; and on/off site production.

Table 15.3 Design process

Blocker	Blocking source	Eliminator	Workplan
Poor project brief	Business analysis and facilities management Inexperienced client	Comprehensive brief	Standard brief information requirements Code of practice for client advice **KBS and CBR support systems to the briefing process**
Budget changes	Brief changes Interaction between design and construction	Design freeze at set stages Buildability prototyping	Project process with staged deliverables Process teamwork
Coordination of design team	Project execution Intellectual ownership Contractual boundaries	Team-based design Shared ownership Different contract	Process teamwork **IT teamwork enablers** Contractual support for process teams
Starting each design from scratch	Briefing	Use of good previous cases	**CBR support systems and industry KBS**
Built-in multi-disciplinary obstacles	Project definition	United design activities	Best practice studies of an integrated design process
Poor design stage transition	Project definition	Allocation of responsibility for design transition management	Best practice design transition management in other sectors Guidelines of design transition management
Need for staged design presentation	Fee system and defined sub-process	Definition of payment and process performance on overall value contribution rather than discrete sub-process optimisation	Redefined project plan of work Redefined consultant fee staged payments guidelines

Lack of formal review and post-occupancy evaluation	Procurement and project definition	Defined design review and evaluation activities	Research into design review methodologies Best practice design review from parallel sectors
Lack of as-built drawings or building model	Design, briefing, facilities management	Building design capture upon completion	**Technology development into modelling systems** Specification of as-built model as design deliverable
Failure to pass on building model	Project definition	Sub-process building model capture and sharing	Specification of building model as intermediate design stage deliverable
Inadequate knowledge sharing, capture and interrelation	Design management	Capture and sharing of design intent	**Development of KBS design engineering tools**
Lost design decisions	Design management	Integration of design	**Research into intelligent integration of information**
Constructability	Design and construction interface	Standardisation, team design Prototyping	Standard designs and components Process teamwork **CAD and VR prototyping**

Items in bold are specific IT research topics.

Current use of IT in the design process

Again, work described earlier in best practice IT benchmarking allows us to give a good comparative picture (Construct IT, 1996b). Being such a design-driven process, construction has traditionally invested heavily in, and pioneered, CAD technology. However, architectural design is a highly creative process. IT use in detailed and scheme design stages closely approximates to world-class practice found in car manufacture. Concept and brief design IT support is relatively poor in construction.

Table 15.4 Project definition process

Blocker	Blocking source	Eliminator	Workplan
Inexperienced client	Business analysis and facilities management	Single proxy client Learn from experienced clients	Clients acting for others in a consultancy basis Client benchmarking workshops
No single client focus	Business analysis and facilities management	Appoint single client representative	Code of practice for briefing – set best practice
Single contact point consultant	Project definition	Appoint consultant representative	Consultant process teams with single point of responsibility and contact
Poor conversion of brief into definition	Business analysis and facilities management	Improve business analysis Standard information requirements	Code of practice for briefing Benchmark best practice **IT tools: CBR and KBS**
Unattributed brief responsibility	Business analysis and facilities management	Single point client responsibility	Code of practice Establish single point responsibility
Sequential consultant working	Project definition	Concurrent engineering	**Develop IT coordination tools** Study pilot projects

Items in bold are specific IT research topics.

Current use of IT in the project definition process

Chapter 11 on benchmarking also described current IT support to briefing in construction which closely approximates to project definition. In comparison to world-class practice in car manufacture, we found that IT support at the early stages of briefing was relatively strong but very weak in the transitional period handing over to design.

Table 15.5 Business analysis and facilities management process

Blocker	Blocking source	Eliminator	Workplan
Inexperienced client	Client	Proxy client Standard information requirements VR prototype	Guidebook for best practice information requirements **VR simulation systems**
Lack of single point responsibility	Client	Project manager Virtual organisation structure	Research into how a project virtual organisation works Contractual virtual organisation support
Lack of facilities management co-ordination	Client	Appoint facilities manager	Guidebook of best practice facilities management advice
Poor industry-wide data	Whole process	Availability of relevant information	Research into means of capturing data and knowledge at source on • economic performance and context • industrial structure and operation using • **KBS and CBR technologies**

Items in bold are specific IT research topics.

Current use of IT in the business analysis and facilities management process

A benchmarking study (Construct IT, 1997c) has considered facilities management. In comparison with world class practice in power plant operation, construction uses IT relatively poorly. It uses IT more effectively in the areas of building services management, environmental management, domestic services and strategy. Relatively poor performance is found in the area of administration and support services.

Table 15.6 Evaluation process

Blocker	Blocking source	Eliminator	Workplan
Poor information to evaluate	All processes	Structure information systems Set project deliverables	**Research IS use in construction**
No formal evaluation procedures	Project definition	Identify review process	Defined evaluation procedures Establish best practice
Poor use of visualisation technologies	Design	Development required	**Research use of CAD and VR technologies**
Single project responsibility	All processes	Set single-point evaluation responsibility	Contractual support for single point responsibility

Items in bold are specific IT research topics.

Current use of IT in the evaluation process

This is an area where best-practice construction IT benchmarking is yet to be undertaken. Our view is thus more subjective. Much previous research work has developed into simulation modelling, little of which has been taken into widespread industry practice.

Table 15.7 Project execution process

Blocker	Blocking source	Eliminator	Workplan
Project coordination	Contractual boundaries Intellectual ownership Lack of teamwork incentives Divided responsibilities	Allocation of virtual organisation responsibilities Motivate to work in teams Shared information ownership	Set up formal virtual organisation structure Process contracts Benchmarking and best practice Process teamwork Motivational issues Remove contractual and intellectual boundaries
Sequential working	Procurement routes Coordination of parties	Concurrent engineering IT coordination	Reset concurrent working – benchmark **Research group coordination technologies**
Poor information for progress tracking	Disparate parties, contractually separated	IT tracking methods Deliverables for information	**Research tracking technologies** Set information requirements

Items in bold are specific IT research topics.

Current use of IT in the project execution process

Again, no specific benchmarking has been undertaken to date related to this issue. However, insights from benchmarks in other areas suggest that group coordination technologies that are widely available and are being used in other industries are not being widely exploited in construction. There is enormous scope for video and desktop conferencing to be widely used given the way that construction has always been a pioneering form of virtual organisation.

Table 15.8 Procurement process

Blocker	Blocking source	Eliminator	Workplan
Inexperienced client	Misunderstanding project Poor industry advice	Standard information requirements Better information of procurement routes	Procurement code of practice **Procurement support through KBS and CBR**
Brief information	Project definition	Standard information requirements	Procurement code of practice **Procurement support through KBS and CBR**
Fragmentation of project team and temporary project structure	Standard contracts	Virtual organisations Process contracts	Standard contracts to address virtual organisation and process structure
Late appointments	Poor planning	Define virtual organisation structure Partnering Design and build and multi-disciplinary design Improved team work Appointments at set stages	Procurement support of virtual organisation Research into partnering Best-practice procurement integration Multi-functional workshops Capturing experience of working Research easier means of fast-tracking relationships
Supply chain management	Project execution	Effective planning	Deliverables as part of procurement

Items in bold are specific IT research topics.

Current use of IT in the procurement process

The first Construct IT benchmarking study into supplier management (Construct IT, 1995a) and its subsequent update (Construct IT, 1998a) approximate closely to the procurement process. They showed construction IT use generally to be far from the world-class best-practice to be found in car parts supply. Selecting suppliers, handling enquiries and managing contracts are areas where construction practice is relatively strong. Subjects of communications and sharing of information on requirements are particularly weak.

Table 15.9 Cost management process

Blocker	Blocking source	Eliminator	Workplan
Design changes Sequential consultant working Late consultant and contractor appointments Lack of industry-wide data	Design management Business analysis and facilities management Project definition Incomplete design Late design	Design freeze Defined deliverables Clear brief Concurrent working Process teams Standard designs	Set stage deliverables **Structured information systems to inhibit contingencies** Process teamwork **CBR and KBS design solutions** Design libraries
Brief changes	Inexperienced client Project definition	Comprehensive briefing Defined deliverables	Standard information requirements Clear definitions of information between actors
Cost growth	Economic cycles Risk Claims Cost planning information	Forecasting methods Risk management Procurement changes	Construction forecasting research Risk analysis techniques Non-adversarial contract **CBR and KBS cost planning**
Mono-function design practice	Design management	Increased multi-disciplinary design and design and build practice	Benchmark and disseminate best multi-disciplinary and design and build practice
Value	Design management	Value management	Value management in construction **Intelligent links between CAD and cost databases**

Items in bold are specific IT research topics.

Current use of IT in the cost management process

This activity approximates to a benchmark by Construct IT in Cost and Change Management (Construct IT, 1998b). In this we found that estimating and accounting are areas where IT penetrated earliest and deepest. Systems based on cost modelling and knowledge-based approaches resulting from research have made some impact into practice.

of solutions necessary to achieve the critical business success factors in the core and support processes. The workplan for industry change covers a wide variety of activities. These include: the transfer of process-based methods in other industries; the use of cross-disciplinary workshops; the implementation of benchmarking activities and dissemination of best practice; the development of key technologies; the adoption of industry standards; the implementation of pilot studies; and the commissioning of numerous research projects.

16 Long-term IT research priorities

Martin Betts, Rachael Luck and Denny McGeorge

"There is real potential, I believe, to achieve greater integration of these processes through the more effective use of IT."
– Operations director, Spanish construction consultants

Technology development of a more long-term nature than that envisaged by Chapter 15 is notoriously difficult to forecast. However, it is possible to look at dominant challenges that industry faces at present and to predict some of the types of technology that might emerge to meet them. It is then possible to anticipate the sort of IT tools that such research might lead to and the scenario for how these might be used to improve future project processes. This chapter does this and identifies generic technologies of visualisation, intelligence, communications and integration as the key themes for the future.

- The key challenges being faced by the industry, in the view of the research community, are the separation of design from construction, and poor information flow.
- The key technology themes to respond to these and other challenges are visualisation and integration.
- Specific technologies that are research priorities at present are VR, CAD, KBS and integrated databases.
- The IT tools most likely to emerge are for building performance evaluation, design evaluation, information exchange, concurrent engineering and design visualisation.

"What I really want is to be able to exchange data with our subsidiary companies and foreign locations. The Internet will enable us to improve some kinds of data exchange but I'm sure we need to go even further."
– General manager, German contractor

16.1 Introduction

In this introduction, it is important to clarify the intention and objectives within this chapter. As we shall see in Section 16.4.5, many attempts have been made recently to identify what research should be done from what might be considered a research demand-side, or industry view. That is, what industry and government feel is needed to benefit companies and UK international competitiveness and performance. Innovation only happens when a need is met by a supply. This study is intended to be a supply-side view from the research community of what academics feel could be done in construction IT research. This has been elicited through responses to previously recorded challenges from industry. The intention overall is to show the extent of match between the supply and demand-side views so that enthusiastic and motivated synergy can be encouraged.

This chapter seeks to develop some measure of academic consensus from the UK construction IT research community. Such consensus is very difficult to develop. Normal mechanisms within the academic world, such as conference series and journals, do so only very indirectly, in an unstructured way and with long time-delays between stages in the process. This chapter seeks to add structure and speed to this process. It is an unnatural activity within the academic community and is sufficiently controversial to the principles of academic freedom, unconstrained personal enquiry, and research competition between individuals and institutions to be bound to meet with resistance. That is not a reason to avoid attempting it.

16.1.1 Previous work

Before extending into a description of the methodology and results of the study and its implications for the future, it is important to position the work with regard to its relationship to other studies. There is a feeling of both innovation and strategy-fatigue developing within both industry and academia at the time that this study is being concluded. Without clarifying how this work adds to and consolidates previous studies, there is a danger of fatigue cracks appearing.

Previous work related to this study includes the IT 2005 study from the Construction IT Forum (Construction IT Forum, 1995), the *Technology Foresight: Report of the Construction Sector Panel* (Office of

Science and Technology, 1995), the *Bridging the Gap* report (DoE, 1995), its subsequent feasibility studies and implementation plans (DoE, 1996a–d), the process-based research workplan report from the Construct IT Centre of Excellence (Construct IT, 1995b), and ongoing work arising from the CRISP IT and Process theme groups.

The IT 2005 report (Construction IT Forum, 1995) documented the views of around 30 experts of how some specific technologies would develop up to the year 2005. It was an open-ended set of expositions of future technologies. Its strength was in allowing open discussion of particular future IT issues. It did not set out to draw consensus, analyse implications in a structured way or relate back to emerging demand-side strategies. This chapter of the book differs in these ways.

The *Technology Foresight* study for construction (Office of Science and Technology, 1995) was not particularly related to IT. There was a separate *Technology Foresight* study more specifically on general IT developments rather than their application to industries like construction. The construction foresight study sought to develop consensus from experts of what areas of technology would offer competitive advantage to UK companies over an extended, long-term timescale. It did not set out to assess or relate research supply to defined research need.

The *Bridging the Gap* strategy (DoE, 1995), and subsequent implementation plans and feasibility studies (DoE, 1996a–d), are the most relevant documents to this work. They have directly sought to establish what research needs to be done, what visions need to be achieved, whether they are feasible and how research funding can realise them. They have not sought to establish whether the academic community is in a position to provide the answers. This chapter will identify the extent to which researchers are able to and inclined to respond to the *Bridging the Gap* challenge. As such this work directly seeks to contribute towards the DoE strategy by encouraging a match between research supply and demand within the existing *Bridging the Gap* agenda.

All of these previous reports, and many other studies of the industry that take a futural perspective, were used as the starting point for this study. The 12 industry challenges have been distilled from the general conclusions from these and a large number of other previous studies, which were analysed in the early stages of the work undertaken in preparing this chapter. The debate about realisable research priorities is an ongoing and continuous one. This chapter is the latest instalment in that debate which views the supply side for the first time and directly builds upon all previous work. It is hoped that building on and moving forward the debate relieves some of the strategy fatigue rather than adding to it.

16.2 Research methodology

This project was lead by a steering committee of senior academics. This Committee defined the objectives of the project so that the findings would complement other studies of construction IT, which were reviewed as background. The committee was also instrumental in developing the research methodology used in this chapter. It was formed from single, senior representatives of each of the eight universities and research institutes that formed the research membership of Construct IT at the time that the study commenced.

The twelve-month, part-time research project was conducted in distinct stages allowing the findings from one stage to inform the structured collection of data and analysis for the next. The framework of the methodology has been represented diagrammatically in Figure 16.1 to show the distinct stages and the sequence in which they occurred.

16.2.1 The research process

A basic premise underpinning the study was that there are issues/challenges facing the construction industry which may be alleviated through IT. This assumption has allowed us to develop a series of statements of challenges facing the construction industry and to ask an academic audience to comment on what future IT research is needed to address each statement. As such, this process has elicited collective views of a substantial sample of the UK construction IT research community as to what research should be conducted to address current and future industry needs.

16.2.2 Review of recent reports of the industry

The first stage of the project was given a head start by being able to build upon the CIOB's Research Agenda Setting Exercise in which members of the research team had been involved. The summaries of reports reviewed for that exercise were augmented with more recent studies of the industry and construction IT reports. This

Figure 16.1 Research methodology

Review of recent reports of the industry
A series of recent, critical industry reports were reviewed to generate a list of the major challenges facing the UK construction industry at the end of the twentieth century.

↓

Decision rules
A series of decision rules was established to select from the long list of major challenges to arrive at just 12 statements of challenges facing the UK construction industry.

↓

Questionnaire design

Quantitative
Quantitative data were targeted to rank the academic community's view of the priority of each challenge.

Qualitative
Qualitative data were collected to establish academics' opinions of future research needed to address each of the major challenges on the selected list.

↓

Co-nomination process
The steering committee of eight experts each nominated five academic experts to complete the questionnaire. The diverse composition and research interests of the steering committee were amplified in the variety of responses and perspectives provided by the co-nominated pool of 40 experts.

↓

Analysis of the questionnaire

Quantitative
Quantitative data were analysed using statistical techniques including an f-test and t-test.

Qualitative
Qualitative data were analysed using an abridged combination of content analysis and grounded theory to group the findings for each challenge.

↓

Report
This report presents academics' perceptions of the priority of each challenge and their views on the future construction IT research needed to address these challenges, to improve the performance of the UK construction industry.

exercise generated many hundreds of statements of issues and problems, which, for the purposes of this study, we have termed challenges currently facing the UK construction industry.

16.2.3 Decision rules

The challenges identified through the review process were classified and grouped according to a series of pre-defined decision rules, dependent upon the statement's link to IT. This dramatically reduced the number of statements to potentially include within the study. The decision was taken to limit the number of statements within the questionnaire to 12.

Level	Decision rule
(0)	'Source document mentions communication or information flow/management/transfer.' The 'solution' to this 'challenge' may be IT or an organisational or structural change within the industry.
(1)	'Source document mentions IT as a generic heading.' This level may include IT technologies which need development, or technologies which are yet to be developed, i.e. generic, 'blue skies' research.
(2)	'Source document mentions an area of IT, e.g. CAD, CAM, etc.' These statements mention a particular technology which needs either developing or applying to the construction industry.
(3)	'Researchers' interpretation of a potential linkage to IT.' The researchers identify statements which have a potential linkage to IT. For this interpretation to be possible the IT application will draw upon an existing or established technology. These statements will identify areas where future research is needed, and, at the near market end, where software needs to be developed.
(4)	'Researchers' interpretation of an industry challenge which could be automated.' These statements cite specific research projects, e.g. industry-wide databases which should be established.

We decided to establish Level 4 as the cut-off point of statements to include within the questionnaire.

The higher order headings, 0 and 1, included statements of a more abstracted and potentially 'blue skies' IT research. The statements within lower order headings, 2 and 3, drew upon existing IT technologies and their transfer into the construction industry. The statements within Level 3 are particularly interesting and relevant to the study. Level 4 statements are specific, their IT use may be secondary to the challenge facing the industry, e.g.

using data-loggers and micro-processors making data more accessible. Some of the Level 4 projects may already be underway.

An attempt has been made to rationalise the reduction of the number of statements in the questionnaire. This was done by eliminating duplication, statements which had a similar content. Other statements which seemed less pertinent to the objectives of the project were similarly removed.

The 12 trigger statements were not intended to give exhaustive coverage of the challenges facing the construction industry but to represent a broad spectrum of interests and to prompt response to the challenge from different academic perspectives. In this way the responses would be unbound and reflect diverse opinion.

16.2.4 Questionnaire design

A simply designed questionnaire was used to collect views from the academic community. The respondents were simply asked to suggest future areas for research to address each challenge. A qualitative approach, asking them to describe these in their own terms, was used because we did not want to influence or constrain the issues. The intention was to establish opinions on the contribution that IT can make to address the burning issues of today's construction industry.

16.2.5 Co-nomination process

The questionnaire was sent to five academic experts nominated by each member of the steering committee. As the members of the steering committee itself had been nominated by their institutions, they had diverse characteristics in terms of expertise and discipline backgrounds. Their nominees would similarly draw upon expertise from a broad spectrum of knowledge. This two-stage collection of data was intended to allow a broad representation of opinion, as specialists in particular areas may be nominated in either stage of the process.

16.2.6 Analysis of the questionnaire

Two methods for analysing qualitative data were considered – grounded theory and content analysis – but both were thought too unwieldy in isolation given the volume of data and time limitations.

Grounded theory, 'the discovery of theory from data', allows the analysis of data from the data itself and groups clusters of similar findings. This approach seemed partially appropriate as it avoided having preconceived ideas of how the data would be grouped.

Content analysis involves analysing a text by breaking it down into discrete sections. The section is then assigned a code or keyword linking the text to an abstract element, which is used to group responses. This approach was pertinent as unbound, textural data were collected.

For a fuller discussion of issues relating to methodologies for qualitative research, the reader is referred to Glaser and Strauss (1967), Turner (1983) and Krippendorf (1980).

A combination of both approaches was used to draw upon the richness of qualitative data gathered from experts and to group the responses within each question. The responses to each question have been read and re-read, grouping snippets of data together on a similar subject (this has similarities with a grounded theory approach). This approach allows the data to inform and determine the categories.

The full report, on which this chapter is based, is structured into 12 sections, according to each of the challenges facing the industry. For each challenge we have written a narrative which ties together the comments and observations made. A conscious effort has been made to reflect the variety of perspectives and interpretations, allowing the diverse and sometimes conflicting views of academics to be presented.

The sections in the main report begin with the issues which were raised because of differing academics' views. This sometimes includes direct quotations of what individual leading academics see as their intended response to an industry challenge. The second section is a classification of generic technologies used as a framework to position suggestions of new technologies that academics feel should be explored in relation to each challenge. Specific tools that may emerge through using these technologies are then listed.

Each section concludes with a summary of future research the academic community is proposing to address that challenge. This is the heart of the research agenda that the academic community feels is its medium-term research strategy, in response to current industry challenges, taking note of interests, expertise and current academic plans.

Finally, each section's major issues and suggestions are summarised as a short scenario. Scenarios are widely advocated as useful strategic planning mechanisms. It is hoped that these summary descriptions of possible future situations may help inform the future IT investments and plans of organisations. These full data and results are not repeated in full in this chapter. Instead we have summarised the findings and their implications for the way construction IT is changing and will change construction business. We have also reproduced the full data for the two challenges perceived as most important by the research community.

16.3 Research responses to industry challenges

The detail of the study behind this chapter contains much information about issues raised, technologies that will emerge, tools that may result, research themes, and an IT scenario for the future. Space does not permit this detail to be reproduced here in full. The challenges cover a range of business, social, environmental, political and legal issues concerned with construction processes and products. In the context of the rest of this book, it is worth noting that the broader research community shares the views we present elsewhere in this book, namely that integration and process change are at the heart of industry improvement needs. For this reason, the full detail of the analysis for these two challenges is repeated here.

16.3.1 Industry Challenge 4

The separation of the processes of design, construction and operation of buildings is a major challenge facing the construction industry.

Views on the importance of this area of research

This is a major problem, which is hindering the construction industry. The fact that this problem has been known and discussed for years did not reduce its relevance. There was wide agreement that this problem must be addressed. The breakdown of the barriers of design, construction and operation is very important to the success of a project and the running of a building.

It is debatable whether these processes should be separated; the construction process should always be integrated. The view was expressed that these processes cannot be anything but separate, but their integration, in terms of the exchange of information between them, has a high priority.

Only one respondent didn't consider this to be an issue for future research and thought that the litigious culture of the industry, which depends upon the divisions of responsibility for these processes, was the root cause, rather than the separation of design and construction processes.

Issues raised by this challenge

Much research in the past has focused on integrating these processes by linking together software applications. Whilst this has been shown to be possible, further development of tools and techniques is necessary to enable this to be carried out in a much more flexible and creative way.

A counter view was posed that maybe IT cannot help with this challenge. Breaking down the barriers between design, construction and operation has been partly addressed in recent years by the adoption of project management teams, i.e. an organisational and process solution. This observation emphasises that the challenge is not just an IT issue.

It was observed that much work is already underway in this area. Major efforts have been undertaken in the last few years to sort out this problem. These tended to be technological solutions but it was acknowledged that the process needs to be refined first.

Technologies

Table 16.1 Technologies relevant to Challenge 4

Visualisation	Intelligence	Communications	Integration
Multimedia Buildability VR models	Buildability KBS	Data communications and exchange standards	Product modelling Data integration standards Object modelling

Tools

Tools are needed to:

- integrate organisations;
- exchange information; and
- facilitate communication.

Virtual organisations are likely to combine all these issues together.

Future research

The suggestions for future research have been grouped into four areas: information standards, design standardisation, developing a building model and a review of methods of working within the industry. These suggestions are all concerned with improving information flows throughout the construction team.

Access to information and standards

- Look at modes of communication that facilitate integration.
- Develop data communication standards for the delivery of information.
- Develop data integration standards.
- Develop IT tools for integrating the activities of the participants involved at all stages of a project.

Design standardisation and tools

- Devise standardised components and operational industry standards which are made accessible via IT tools.
- Component-based building, rationalised design, manufacture to world-class standards, faster, better quality building.
- Use knowledge-based systems to advise on practicality and constructability issues.

Change the construction process

- Research new methods of working which break with accepted roles of client, architect, contractor and subcontractor. Adopt partnering and team-working in technical clusters. IT can assist by allowing collaborative working using multimedia.

- Understand the integration issues by looking at the actors in the process.

Building model

- Develop an integrated building model.
- Develop an industry-wide product model which evolves throughout the design stage and is updated during construction and post-occupancy. This will assist the efficient flow of information at the interfaces.
- Develop an 'electronic building system' that allows visualisation of the building (VR) and investigates its performance from many perspectives (energy, people movement, demolition, etc.).

Future scenario

Within the next ten years, the construction industry in the UK will achieve much greater integration of the processes of design, construction and facilities management. The construction and future operation of buildings will be assessable at design stage through integrated IT, based on established information and design standards.

16.3.2 Industry Challenge 11

The flow of information amongst project participants is currently not coherent or comprehensive.

Views on the importance of this area of research

This statement was ranked the highest in terms of its relevance to the performance of the construction industry. Several of the respondents thought that there were similarities between this challenge, which is concerned with information flow, and Challenge 4, which looks at the separation of the construction process into different stages. Challenge 4 was ranked the second highest and the two together were shown through statistical tests to be distinct from all other statements in being clearly more relevant to the research community.

This was considered to be a major issue facing the industry that has largely been ignored by research funding initiatives. Most respondents considered this to be a very important topic but thought that it was the gist of a lot of IT work at the moment.

Issues raised by the challenge

The construction process is highly dependent upon the transfer and exchange of information between the project participants. It was considered that the flow of information was a key industry issue as the effectiveness of the transfer could bring about the integration of the project stages. IT was considered to be a relevant approach for addressing this challenge as it would allow the process to be reviewed and the medium of exchange to be adjusted.

A group of respondents thought that there was a need to better understand the information requirements within the construction process if real improvements in construction performance are to be achieved. Much initial research in data flow modelling within organisations has been carried out. This work would act as a precursor to the development of IT systems.

It was considered that improving the information flow within current practices would only have a temporary effect. For significant improvements this problem must be addressed within a completely redesigned integrated process environment.

Those who considered it important that information at various levels within a project is handled consistently reiterated this view. Each person within a project may need information for different purposes; therefore making it accessible, in a consistent database, is a very difficult task and worthy of a major piece of work.

The observation was made that even if the flow of information were perfectly coherent and comprehensive, it would not *per se* guarantee better design or project management or an improved product. In this respect the research response to the challenge would enhance the construction process but not necessarily the physical construction product.

Technologies

Table 16.2 Technologies relevant to Challenge 11

Visualisation	Intelligence	Communications	Integration
VR models	Knowledge management	The Internet and intranets	Product modelling
		On-line product databases	Object-oriented programming
			Integrated project database

Tools

- Knowledge handling tools;
- on-line manufacturers' databases.

Future research

This challenge generated the most suggestions for future research. These have been grouped into three categories: studies of the use of IT and information flows, IT to improve the design process and suggestions for IT research which will be of benefit throughout the construction process.

Studies of the use of IT

The use of IT should be studied as a socio/political issue as well as a technological one.

- Investigate the cultural implications of an integrated project environment for all parties involved in the construction process.
- Model existing information flows and redesign.
- Develop means of audit-trailing the sequence of decision making, who did what and when.
- Study how project participants contribute information to a project, how they receive and disseminate information.

IT to improve the design process

- Develop IT methods for user dialogue capture, storage and organisation (blackboarding techniques).
- Develop IT for user dialogue handling (concept mapping techniques).
- Develop new approaches to knowledge handling, which separate out user, application, design process and data knowledge.
- Use buildability VR models in the design environment.
- Develop better construction models for use by designers.

IT to generally improve the construction process

- Develop integrated project databases;
- Develop the technology to deliver an integrated project environment.

- Develop a single product model which is updated throughout the life cycle of a project.
- Develop object technology expertise in construction.
- Develop process and product models using component and distributed computing techniques.
- Innovative use of the Internet and intranets for the distribution and sharing of information.
- Standardise information exchange.
- The integration of hardware and software platforms needs investigating.

Future scenario

Many projects in the future will have greatly improved flows of information between their participants. This will use diverse technologies to work towards the integrated project database concept. This improved information flow will lead to re-engineered and improved processes of design and construction.

16.4 Overall findings

16.4.1 Consensus and diversity

This chapter has sought to identify consensus within the academic community on future research issues in construction IT. In the study behind the writing of this chapter, a very diverse group of experts have been consulted. In these circumstances it is worth noting the degree to which an overall high level of consensus has been established. Clear patterns have emerged in qualitative data that permit us to draw conclusions from the extensive exercise.

Statistical tests on the importance of the challenges divided them into three distinct groups in terms of their priority, which were not due to chance. This rank order of priority is used in some of the tables that follow in this section. There was a clear consensus on where the priorities lie. Beyond consensus, the overall study also demonstrates considerable variety and richness of issues, technologies and visions within the field. This combination of observations has a number of implications for future research funding.

One is that building on success in research programmes and having strong programmed research is necessary in long-term research, medium-term development and short-term innovation. We do tend to have clear (but discontinuous) programmes and themes in short-term innovation funding but not in medium- and long-term research.

The other is that all stages of research, development and innovation have considerable diversity in topics that different sections of the research community see as important. This suggests that all funders should allow some flexibility for creativity and diversity in research funding even within tightly defined research programmes.

16.4.2 Issues raised

The overall issues raised are very diverse as shown in Table 16.3. A consistent message of information flow, transfer and integration and supporting changes in processes is evident. The importance of design processes, and their improved understanding, also emerges from the work. The way that non-IT and interdisciplinary issues impact on the challenges emerges in a number of ways. The other common issue to emerge is the concern with quality of life and building performance.

The implications of this pattern of issues raised are that IT

Table 16.3 Issues raised by challenges

Challenge no.	Issues raised within each challenge
11	• information transfer and exchange • information flow and integration • understanding information requirements • integrated environments • consistent information modelling
4	• distributed information integration/ interoperability • process integration • combined process and data integration
3	• replicating best practice • feedback from previous projects • coping with site peculiarities • relevance to SMEs
8	• design stage constructability input • timeliness of site information • nature of the design process • influence of procurement method

Contd

Table 16.3 *Contd*

Challenge no.	Issues raised within each challenge
7	• organisational boundaries – fragmentation • improved application of existing technologies • poor manual communication culture
2	• interdisciplinary need • better understanding of non-IT process • limited IT potential
12	• building performance prediction tools • quality of life dimension • integrated environment • problems with complete integration in a fragmented industry
1	• importance of environmental issues • macro/micro dimensions • need for societal/cultural change
5	• alternative types of building model
9	• the firm is important and greatly ignored at present
10	• relevance of education • integration of a variety of performance matrices • need for legislation and cultural change • conflict between wealth creation and quality of life
6	• including views of occupants rather than clients

research programmes need to be linked to general process research programmes more closely if they are to realise their full potential. The fact that IT research has a strong interdisciplinary dimension should be recognised. The potential for IT research to impact non-wealth creation dimensions should also be recognised.

16.4.3 Generic technologies

Respondents within this study have cited a number of different technologies. Within the specific analysis of responses to individual challenges, and within this summary here in Tables 16.4 and 16.5, the classification has been used of four generic technologies. The expression *veni, vidi, vici* refers to coming to a problem, seeing and understanding it, and seeking to solve it. Within Construct IT we are progressing through a process of bringing relevant organisations to a shared innovation community, seeing and understanding the problem through benchmarking current industry practice and setting national strategy, and promoting research into generic technologies as a means of solving current practice and future vision needs.

For the VICI solutions, the generic technologies that are emerging are visualisation, intelligence, communications and integration. The summary Table 16.4 shows a robust classification of technologies that have directly emerged from this exercise. The implications for long-term research funding are that for the UK to be strongly placed to support technological innovation in the long term, significant and continuous programmes of funding are required in the four generic technologies. Each requires careful support and cumulative development if the UK innovation community is to be strategically well positioned to support applied innovation in areas of long-term need.

Of the generic technologies, the research community within this

Table 16.4 Technologies suggested by challenges

	Visualisation	Intelligence	Communications	Integration
1	Virtual reality (VR), other visualisation technologies	Knowledge based systems (KBS)	–	Databases
2	VR, multimedia, 3D models, visualisation	Case-based reasoning (CBR), KBS, knowledge elicitation	–	Databases, modelling information
3	VRML (virtual reality modelling language)	IKBS, CBR, data mining	Site communications, auto identification, barcoding	Data warehousing, information management

Contd

Table 16.4 *Contd*

	Visualisation	Intelligence	Communications	Integration
4	Buildability VR models, multimedia	Buildability KBS	Data communications, exchange standards	Object modelling, product modelling, data integration standards
5	VR, CAD, QuickTime VR, 3D modelling	–	Computer supported collaborative working	Integrated databases, object modelling
6	VR, CAD	Expert systems		Databases
7	–	–	Bar coding, CSCW, Internet, EDI	Concurrent engineering, integrated databases
8	VR CAD	KB simulation, KBS for structural frames	Site communications	Object libraries
9	VR	Intelligent CAD	Automated data capture	Integrated databases
10	VR	Energy modelling	–	–
11	VR	Knowledge management	Internet, intranets, on-line product databases	Product modelling, object-oriented programming, integrated project databases
12	CAD	Intelligent simulation, intelligent materials	–	Integrated engineering

Table 16.5 Summary of generic technologies

	1	2	3	4	5	6	7	8	9	10	11	12	Totals
VR	*	*		*	*	*		*	*	*	*		9
Multimedia		*		*									2
3D models		*			*								2
Visualisation	*	*											2
VRML			*										1
Buildability models				*									1
CAD					*	*		*	*			*	5
QuickTime VR					*								1
Visualisation total													**23**
Knowledge-based systems	*	*	*	*		*		*	*				7
Case-based reasoning		*	*										2
Data mining			*										1
Knowledge management											*		1
Intelligent materials												*	1
Energy modelling										*			1
Knowledge elicitation		*											1
Intelligence total													**14**
Auto identification			*										1
Barcoding			*				*						2
Site communications			*					*					2
Data communications				*									1
Exchange standards				*									1
CSCW					*								1
Automated data capture									*				1

Contd

Table 16.5 *Contd*

	1	2	3	4	5	6	7	8	9	10	11	12	Totals
The Internet							*				*		2
Intranets											*		1
On-line product databases											*		1
EDI							*						1
Communications total													**14**
Databases	*	*				*							3
Object modelling		*		*	*								3
Integrated databases					*	*	*		*		*	*	6
Product modelling				*							*		2
Object-oriented programming											*		1
Data warehousing			*										1
Information management			*										1
Concurrent engineering							*						1
Data integration standards				*									1
Object libraries								*					1
Modelling information		*											1
Integration total													**21**

study has made most reference to technologies of visualisation and integration. As generic technologies, it appears that these are perceived as the key generic technologies to construction at present. Within the generic types of technologies, specific technologies that are identified as priorities by the research community are virtual reality, knowledge-based systems, CAD and integrated databases.

16.4.4 Tools

Although not the specific domain of the academic community, the respondents to this study have suggested a number of tools that may emerge from future research. This list of items may be most relevant to an industrial audience for this chapter. Within Tables 16.6 and 16.7 the importance of design tools and integration tools is clearly evident. However, the most commonly cited tools are those concerned with building performance. This is a way in which the research community's agenda can be shown to extend beyond a narrow focus with wealth creation.

Table 16.6 Tools suggested by challenges

Challenge no.	Tools considered necessary to address each challenge
11	• information exchange • organisational integration • communication
4	• information exchange • organisational integration • communication
3	• site/office links – shared working environment • CAD/management integration • site planning and management • auto identification/barcoding
8	• constructability simulation • site construction information • cost simulation • buildability checklist
7	• the Internet • EDI • concurrent engineering • collaborative communities
2	• on-line collaboration • simulation models • tools to capture client's requirements • auto specification • cognitive stakeholder mapping

Contd

Table 16.6 *Contd*

Challenge no.	Tools considered necessary to address each challenge
12	• computational fluid dynamics simulation tools • CAD links with performance prediction • KBS for materials performance
1	• GIS tools • life-cycle cost tools • embodied energy prediction tools • energy consumption prediction tools • visual impact assessment • impact of traffic flow • on-air collaboration • building performance evaluation
5	• VR design modelling • 3D models for visualisation and structural analysis • as-built models • construction process archiving • full life-cycle models • CAD/FM integration
9	• building performance simulations
10	• embodied energy simulation • project audit checklist • environmental visualisation
6	• simulation tools

16.4.5 Research themes

Each challenge's responses have resulted in a classification of some particular research themes that emerge from the analysis. Within Tables 16.8 and 16.9, these research themes are summarised. One dominant research issue is improving the access of construction industry participants to information. The other dominant research theme is concerned with improving or reviewing industry processes. Secondary research themes are design and its evaluation and the macro-industry impact of IT innovation.

Table 16.7 Summary of tools

	1	2	3	4	5	6	7	8	9	10	11	12	Totals
Information exchange within project team				*			*				*		3
Databases of product/ project information								*				*	2
Concurrent working	*		*				*						3
Tools to integrate packages			*		*								2
Decision mapping tools		*											1
Design visualisation tools		*			*					*			3
Design evaluation tools		*			*			*				*	4
Building performance evaluation tools	*					*			*	*		*	5
Cost simulation	*							*					2
Planning tools			*										1
Auto identification tools			*										1
Building model						*				*			2

16.4.6 Overall findings

In summary, the overall findings of the study behind this chapter can be summarised as follows.

- A high level of consensus exists, within members of the UK academic community for construction IT, of research priorities for the future.
- The dominant challenges facing UK industry in the view of the academic community are those of integration of the industry and its processes.
- Key issues within the research community embrace the search

Table 16.8 Research themes suggested by challenges

Challenge no.	Research issues
1	• materials and components • site processes • macro-environmental impact
2	• access to information • identifying client needs • visualising designs
3	• process review • document control
4	• access to information and standards • design standardisation and tools • refining the construction process • building model
5	• object models • building models • interfacing models
6	• measuring performance of buildings in use • data capture techniques • building models • performance prediction tools • control loops for buildings in use
7	• studies of the process • technological developments to improve information flow
8	• access to information • design evaluation • longer term changes to the industry
9	• design evaluation tools • access to information
10	• materials selection • design evaluation tools
11	• studies of the use of IT • IT to improve the design process • IT for the benefit of the industry as a whole
12	• simulate and model designs • better understand the design process

Table 16.9 Summary of research themes

	1	2	3	4	5	6	7	8	9	10	11	12	Totals
Identifying client needs		*											1
Access to information		*	*	*		*	*	*	*				7
Visualising designs		*										*	2
Design standardisation and tools				*									1
Design evaluation								*	*	*			3
Materials selection	*									*			2
Building models					*	*							2
Process review	*		*	*				*			*	*	6
Performance of buildings in use							*						1
Studies of the use of IT											*		1
Macro industry impact	*							*			*		3

for integration, the need for interdisciplinary research and a concern with the non-business issues of quality of life.

- The academic community recognises clear generic technologies of visualisation, intelligence, communications and integration. All generic technologies need to be supported on a continuous basis for the long term in the strategic interests of the UK construction industry.
- Visualisation and integration are considered by the academic community to be the key generic technologies of strategic significance to UK construction.
- Virtual reality, knowledge-based systems, CAD and integrated databases are considered to be the key specific technologies to UK construction.
- Key tools that are likely to emerge in the future from academic research and its subsequent development in industry are concerned with building performance, design and integration.
- Key research themes within the academic community include access to information and the review of the industry's processes.

Table 16.10 Overall summary

Tools	6	10	9	5	1	12	2	7	8	3	4	11	Totals
Information exchange within project team								*			*	*	3
Databases of product/ project information						*			*				2
Concurrent working					*			*		*			3
Tools to integrate packages				*						*			2
Decision mapping tools							*						1
Design visualisation tools		*		*			*						3
Design evaluation tools				*		*	*		*				4
Building performance evaluation tools	*	*	*		*	*							5
Cost simulation					*				*				2
Planning tools										*			1
Auto identification tools										*			1
Building model	*	*											2

Technologies	6	10	9	5	1	12	2	7	8	3	4	11	Totals
VR	*	*	*	*	*		*		*		*	*	9
Multimedia							*				*		2
3D models				*			*						2
Visualisation					*		*						2
VRML										*			1
Buildability models											*		1
CAD	*		*	*		*			*				5

Contd

Table 16.10 *Contd*

Technologies	6	10	9	5	1	12	2	7	8	3	4	11	Totals
QuickTime VR				*									1
Visualisation total													**23**
Knowledge-based systems	*		*		*		*		*	*	*		7
Case-based reasoning							*			*			2
Data mining										*			1
Knowledge management												*	1
Intelligent materials						*							1
Energy modelling		*											1
Knowledge elicitation							*						1
Intelligence total													**14**
Auto identification										*			1
Barcoding								*		*			2
Site communications									*	*			2
Data communications											*		1
Exchange standards											*		1
CSCW				*									1
Automated data capture			*										1
The Internet								*				*	2
Intranets												*	1
On-line product databases												*	1
EDI								*					1
Communications total													**14**

Contd

Table 16.10 *Contd*

Technologies	6	10	9	5	1	12	2	7	8	3	4	11	Totals
Databases	*				*		*						3
Object modelling				*			*				*		3
Integrated databases	*		*	*		*		*				*	6
Product modelling											*	*	2
Object-oriented programming												*	1
Data warehousing										*			1
Information management										*			1
Concurrent engineering								*					1
Data integration standards											*		1
Object libraries									*				1
Modelling information							*						1
Integration total													**21**
Research beyond 2000	6	10	9	5	1	12	2	7	8	3	4	11	Totals
Identifying client needs							*						1
Access to information	*		*				*	*	*	*	*		7
Visualising designs						*	*						2
Design standardisation and tools											*		1
Design evaluation		*	*						*				3
Materials section		*			*								2
Building models	*			*									2

Contd

Table 16.10 *Contd*

Research beyond 2000	6	10	9	5	1	12	2	7	8	3	4	11	Totals
Process review					*	*			*	*	*	*	6
Performance of buildings in use								*					1
Studies of the use of IT												*	1
Macro industry impact					*				*			*	3

16.5 Overall future scenario generated by this study

Data for each of the 12 industry challenges resulted in 12 individual scenarios for the future being constructed. The following is an overall summary amalgamation of these and represents a paraphrasing of what leading researchers think the future construction industry will be like when better supported by IT.

The most important way in which the future construction industry in the UK will differ from the present will be through a greater measure of integration. Many projects in the future will have greatly improved flows of information between their participants. This will be achieved through the use of diverse technologies to work towards the integrated project database concept. This improved information flow will lead to re-engineered and improved processes of design and construction. It will also enable much greater integration of the processes of design, construction and facilities management. The construction and future operation of buildings will be assessable at design stage through integrated IT, based on established information and design standards. Beyond this improved integration, the future will also see increased IT support to discrete stages of the overall process.

- One can imagine a scenario within say 5–10 years when a large range of marketable software will be available which has been specifically created for the initial briefing stage of a construction project. This range will collectively allow greater access to information required in briefing, will improve the process of identifying client needs, and will be linked to an effective facility for future design visualisation.
- Building design in the future is likely to be supported by

integrated, intelligent IT applications that have automatically made the details of the design more appropriate to the construction process and give general feedback before it is too late on the overall ease with which the designed facility can be built.

- Sites in the future will be much more IT intensive with the World-Wide Web providing a fast-access opportunity to link to intranets and the Internet. A diverse range of technologies will be combined to allow more effective control of site information as part of a dramatically changed site management process.
- Improved supply chain management will be enabled by IT supporting collaborative work, electronic trading and effective communication between the multiple parties to the construction process. IT will be the trigger to enable construction to approach the high standards of supply chain management already being achieved in other sectors.

The future UK construction industry will also benefit from other changes in the way it works. These might include:

- Projects of the future subject to simulation at their various stages through a range of performance prediction tools. These may link more widely to intelligent simulation of building design. In the future, IT applications may emerge that integrate visualisation and artificial intelligence technology to produce tools that can model the environmental impact of various design and construction decisions. These could include early concept design decisions, material and component selection and decisions related to site processes.
- Projects in the future will be handed over to clients with full object-based models of the building as designed and constructed. These as-built models will have an open architecture and will become critical information for the client to use in managing the facility through its life cycle. As-built models and their further integration with life-cycle data will become key sources of information for future new building designs which seek to learn from the industry's previous experiences.
- The use of IT during facilities management in the future will see a combination of technologies being used to automatically capture data on facilities in use and feed this into design evaluation studies as a basis for simulating future performance.
- Building design in the future will exploit IT to allow 'green' issues to be more adequately addressed. This will include visualisation and modelling technology being used to select

materials and components, and evaluate design alternatives, all in a more environmentally friendly manner.

- Increased IT support to the occupancy phase of projects which will help occupiers better understand the way their building is performing and future designers and constructors to be better placed to learn from other occupiers and earlier lessons.

17 An IT-supported new process

Ghassan Aouad, Rachel Cooper, Mike Kaglioglu and Martin Sexton

"Clients are now not only asking for information on paper, but also for disk-based information. We nearly always have compatibility problems."
– *Project manager, UK*

A project led by key UK industrialists and undertaken by leading academics has developed a new process protocol based on principles from new product development in manufacturing industries. It has activity zones (processes) being undertaken in key project phases. It has been supplemented by a map of IT that could support elements of this new process. It envisages developing maturity in technology and process in parallel.

- A new generic design and construction process could lead to better team-building, improved project stakeholder involvement, better coordination and management, more effective knowledge management and more coordinated IT innovation.
- A new process envisages virtual companies, increased 'front-end' emphasis, progressive design fixity, improved teamwork and communications, single point contact and responsibility, better process and change management, legacy project archives and projects moving through soft and hard gates.
- IT maps of new project processes allow improved process-thinking within IT developments, avoidance of current IT misuse, improved links between business and IT strategies, better understanding of IT capabilities and IT limitations to become more clearly apparent.

"We do the process like this, because we have always done it like this."
– *Head of practice, Swedish engineering consultants*

17.1 Introduction

In Chapters 15 and 16, we have seen what research and innovation priorities in technology must be progressed to allow the industry to follow improved processes. Many of the potential benefits associated with improved processes in construction can be realised with significant IT support. Indeed, IT will only achieve profound change if its introduction and use are linked to changes in the overall conduct of the design and construction process. We have made this point in many of the preceding chapters of this book. This chapter presents an IT map that supports an improved design and construction process. This IT map views IT as an enabler and not as a driver of the process. The work presented in this chapter has been undertaken as part of a major EPSRC-funded project under the IMI (Innovative Manufacturing Initiative) 'Construction as a Manufacturing Process' sector initiative, and in close collaboration with a wide spectrum of construction firms representing clients, consultants, contractors, subcontractors, suppliers and software houses, enabling the consideration of a wide range of views and perspectives. Some case studies related to the use of IT by some of the participating companies are also presented.

Working within a process framework/process map context is becoming the norm in many manufacturing firms. Companies such as GPT, British Aerospace, BAA, BNFL, the Bucknall Group and many other large and small organisations have developed process maps that put them in a good position to meet their customer and business requirements (Kaglioglou *et al.*, 1998a). Undoubtedly, these process maps can help in the development of their equivalent IT maps that position the technologies which enable and support the processes involved within the business environment (Aouad *et al.*, 1998b).

Many studies have been conducted in the construction sector in recent years in order to investigate the relationships between technologies and in particular IT and processes. Most of these studies have concentrated on IT capabilities and forecasting of how IT will be used in the future (Construction IT Forum, 1995; KPMG and CICA, 1993; Aouad *et al.*, 1997). These studies, like Chapter 16, predict the types of technologies that will be used by the industry in the next 10–15 years.

17.2 The generic process protocol

A process protocol can be defined as a way in which the processes involved in the designing and constructing of a structure are arranged so as to produce an efficient, effective and economical way of undertaking the design and construction of projects (Love, 1996; Sanvido, 1992). A process protocol/map should be customisable to meet the needs of any company. Tangible benefits can be realised through wastage reduction, shortening the duration of projects and improving communication methods and channels (Latham, 1994).

Supply chain management in the construction sector can be improved if everyone is working within the same framework. A process map will provide such a framework, which will lead to:

- better team building;
- improved stakeholder involvement;
- better coordination and management;
- efficient knowledge management leading to better learning within organisations; and
- more coordinated IT through IT maps.

This section briefly describes a redesigned design and construction process which covers those stages of construction projects, from conception of the client's needs and feasibility to construction and operation. Essentially, the model breaks down the design and construction process into ten distinct phases. These ten phases are grouped into four broad stages, namely pre-project, pre-construction, construction and post-completion. A simplified picture of the process map (Figure 17.1) is presented.

The pre-project phases relate to the strategic business considerations of any potential project which aims to address a client's need. Throughout the pre-project phases the client's need is progressively defined and assessed with the aim of:

- determining the need for a construction project solution; and
- securing outline financial authority to proceed to the pre-construction phases.

With outline financial approval obtained, the process progresses through to the pre-construction phases, where the defined client's need is developed into an appropriate design solution. Like many conventional models of the design process, the pre-project phases

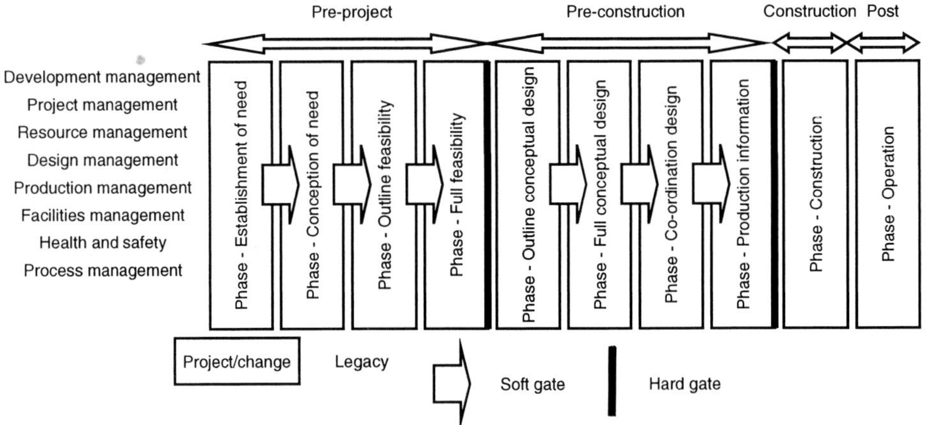

Figure 17.1 A simplified picture of the process protocol

develop the design through a logical sequence, with the aim of delivering approved production information. The phase review process, however, adds the potential for the progressive fixing of the design, together with its concurrent development, within a formal, coordinated framework.

At the end of the pre-construction phases, the aim is to secure full financial authority to proceed. Only upon such authority will the construction phase commence, and this decision will be easier to make where the extent of the works, and the associated risks can be readily understood. The construction phase is solely concerned with the production of the project solution. It is here that the full benefits of the coordination and communication earlier in the process may be fully realised. Potentially, any changes in the client's requirements will be minimal, as the increased cost of change as the design progresses should be fully understood by the time on-site construction work begins.

The hard gate that divides the pre-construction and construction phases should not prevent a work-package approach to construction, and the associated delivery time benefits this brings. As with all activities in the process, where concurrency is possible, it can be accommodated. The hard and soft gates that signify phase reviews merely require that before such an activity is carried out, approval is granted.

Upon completion of the construction phase, the process protocol continues into the post-completion phase which aims to continually monitor and manage the maintenance needs of the constructed facility. Again, the full involvement of facilities management specialists at the earlier stages of the process should make the enactment of such activities less problematic. The need for surveys of the completed property, for example, should be avoided as all records of the development of the facility should have been recorded by the project's legacy archive.

The process map is being developed to support and encourage integration and proper coordination between the various participants or stakeholders of a construction project. The following are key features of this process map:

- the virtual company;
- emphasis on the 'front end';
- design fixity;
- teamwork;
- effective communication;
- single contact point: project director/project board;
- change management;
- process management;
- process and project legacy archive; and
- soft and hard gates.

The virtual company concept involves getting all the stakeholders involved in a construction project to identify their objectives and interests. This helps at the requirements capture phase (Phases 0–1) ensuring that many of the major problems are solved through discussions and most of the requirements are identified by these various stakeholders. This will help in meeting the corporate virtual objectives and shared vision. This is similar to a temporary company structure (Cooper *et al.*, 1998, Kaglioglou *et al.*, 1998a).

The generic design and construction process protocol has also highlighted the need for greater 'front-end' activity in the process (Kaglioglou *et al.*, 1999). Traditionally in the construction industry and unlike manufacturing, the emphasis is on the construction phase where contractual claims arise because of the unconstructable nature of design. The process map presented in this chapter puts the argument that the emphasis on feasibility and design is of crucial importance. In manufacturing, this is the case and IT with its rapid prototyping capabilities plays an important role. Similarly, the construction industry may benefit from the rapid prototyping approach which can help solve most of the problems before construction begins. Virtual reality and 3D modelling tools provide the right medium for rapid prototyping purposes.

Design fixity involves the freezing of a certain phase with the agreement of the stakeholders. For instance, the client can decide on fixing the design phase in order to avoid changes which often prove to be costly. Teamwork and communication are two major aspects of business process redesign. The process map promotes these two aspects through the concept of a virtual company, which results in better relations between the project participants, and through IT-based communication. The single contact point, change and process manager ensures that the process is managed with some authority. Roles, rights and responsibilities are within the scope of the aforementioned concepts. The legacy archive (a library of previous cases and current project data) ensures that best practice is captured and re-used. Finally, soft and hard gates ensure that major decisions are assessed and evaluated. The soft gate implies that decisions could be conditional, in that the project is not stopped for one or two non-critical activities, thus ensuring concurrency and reduced timescales. Hard gates indicate firm and final decisions regarding whether or not to proceed to the next phase within the process. The process map is described fully elsewhere (Cooper *et al.*, 1998).

The concepts outlined above, whilst central to the operation of the process protocol, are not unique to this 'new way of working'. Many of the existing studies, both within construction and manufacturing, acknowledge such concepts as teamwork, design fixity, and stage-gate systems (Cooper *et al.*, 1998). However, if such concepts are to fulfil their ultimate potential, it would appear that IT tools and techniques will play a major role (Brandon and Betts, 1995). The next section of this chapter will describe an IT map which has been developed in order to demonstrate the potential of the use of IT in this context.

17.3 The IT map

The information requirements of the aforementioned redesigned processes need to be modelled (Kaglioglou *et al.*, 1998b). Technology can then be used to enhance integration and sharing of information. Technologies such as object-oriented databases, virtual reality, expert systems, case-based reasoning, neural networks and traditional commercial packages can be used and integrated in order to illustrate these principles and to ensure that information can be shared between the various processes. This is a long-term objective. It will allow the industry to realise the benefits which can be gained from coupling visualisation systems, knowledge-based systems, and traditional commercial packages in support of the redesigned processes. Ideally, the client should have the ability to 'walk through' the designed product using virtual reality based on information stored in the database. The designer should be able to select the most appropriate design using case-based reasoning involving cases where the fabrication processes were simplified. The project planner should be able to establish the project plan using a knowledge-based system coupled with project planning software. As these applications will be linked to an integrated database, this will ensure the integrity of information.

Further to the work presented in the previous chapters, another recent study at the University of Salford has identified the IT topics which will be of crucial importance to the industry in the next 10 years based on a questionnaire survey conducted in academia and industry. A total of 175 questionnaires were sent out into industry, of which 55 replied, a percentage return of 31.4%. The questionnaires were distributed in differing amounts to the three following professions: contractors, quantity surveyors and architects. Samples were chosen from published material on top contractors, quantity surveyors and architects. The return rate of each individual profession was:

- quantity surveyors: 55 sent, 18 replies. Return rate = 33%;
- contractors: 80 sent, 20 replies. Return rate = 25%;
- architects: 40 sent, 17 replies. Return rate = 42%.

In addition, 33 questionnaires were sent to construction IT academics with 14 replies (return rate = 42%).

The results of the questionnaire clearly indicate that communications and networking will be a major topic over the next ten years. This will ultimately give the process some push resulting in some maturity in reaching a consistent, predictable and improved process. The findings of this study have helped in proposing an IT map which reflects the needs and requirements of both industry and academia (Aouad *et al.*, 1997). This survey mirrors and supports many of the findings of the studies into long-term research priorities presented in Chapter 16.

Figure 17.2 clearly shows that topics such as AI (artificial intelligence), neural networks, and simulation are the least important for the progress of construction IT in the next ten years. This is mainly attributed to the slow uptake of these technologies by the industry and not to their potential. This chapter argues that most of the aforementioned technologies will be beneficial at some stage of

Quantity surveyors
1 Communications/networking
2 Databases
3 Computer-aided estimating
4 Computer-aided design
5 Data input technology
6 Integration
7 Strategic planning
8 Multimedia
9 Geographical information systems
10 Computer-aided planning
11 Simulation
12 Artificial intelligence
13 Virtual reality
14 CASE tools
15 Neural networks
16 Robotics

Contractors
1 Communications/networking
2 Computer-aided planning
3 Strategic planning
4 Databases
5 Integration
6 Computer-aided estimating
7 Computer-aided design
8 Data input technology
9 Multimedia
10 Geographical information systems
11 Simulation
12 Virtual reality
13 Artificial intelligence
14 CASE tools
15 Neural networks
16 Robotics

Architects
1 Communications/networking
2= Computer-aided design
2= Databases
4 Integration
5 Virtual reality
6 Computer-aided planning
7= Data input technology
7= Strategic planning
9 Simulation
10 Computer-aided estimating
11 Multimedia
12= Artificial intelligence
12= Geographical information systems
14 CASE tools
15 Robotics
16 Neural networks

Academics
1 Communications/networking
2 Computer-aided planning
3 Computer-aided design
4 Databases
5 Data input technology
6 Computer-aided estimating
7 Simulation
8 Multimedia
9 Virtual reality
10 Integration
11 Strategic planning
12 Geographic information systems
13 Artificial intelligence
14 CASE tools
15 Neural networks
16 Robotics

Figure 17.2 IT priority topics (source: Aouad *et al.*, 1998)

the process. It is the coupling and integration of these technologies which will provide the ultimate benefits to the industry. The findings of the survey have helped the authors to identify technologies which can be used to devise the IT map.

The IT solutions proposed in this chapter have been developed in response to the following needs:

- lack of process thinking within an IT map;
- misuse of current IT tools;
- confusion over linking business and IT strategies;

- lack of understanding of IT capabilities; and
- the limitations of the current technology.

The IT map presented here has been developed in relation to the process requirements. It is clearly shown that there are two types of IT which can support the process: the generic IT such as EDI, AI, integrated databases and document management systems which can be used at any phase of the process in order to exchange information and support communication; and the IT specific solution which may be more helpful at a specific phase. For instance, CAD is useful at the design stage whereas computer-aided planning is more appropriate at the construction stage.

The IT map (Figure 17.3) includes some of the technologies specified by industry and academia at the initial phase (pre-project), where simulation, 'what-if?' and economic appraisal tools are most appropriate. Potential project scenarios could be regenerated from an archived library of previous projects and the emphasis should be on creating this library. These could include VR, 3D and other media. Artificial intelligence techniques including case-based reasoning, neural networks, fuzzy logic, genetic

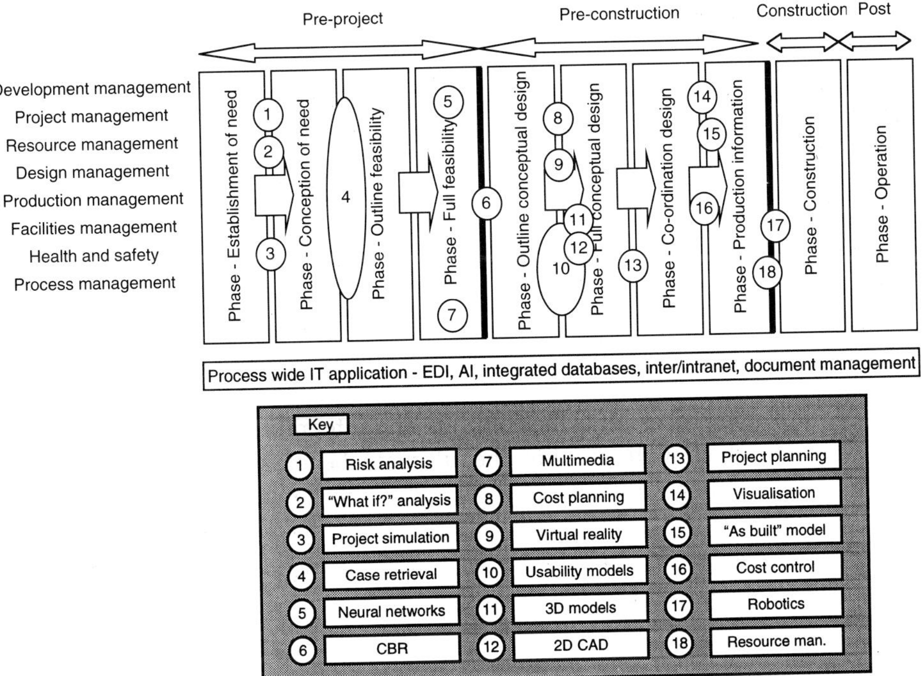

Figure 17.3 The IT map

algorithms and KBS may be appropriate at the initial phase of design where the creativity issue of design plays a major role. Computer-aided design, VR and other tools are not able to capture the information required for this phase, but may prove helpful at the detailed design and construction phases. The IT map clearly shows that there are many tools which can be used to support the various phases of the design and construction process. These tools are generic and can be used by any industry. It is the maturity of using an IT tool within an application context which will help the overall process through a technology push.

A large number of emerging IT tools are considered to be research tools. These will need to have a wider acceptance by the industry through the phases of testing, refining, accepting and finally customising. It is the customisation of the technology for a certain application which will form the starting point for the aforementioned technology push of the process. These issues are the subject of other publications (Hinks *et al.*, 1997; Aouad *et al.*, 1998c).

The four broad stages of the ten phases of the process map described earlier (namely pre-project, pre-construction, construction and post-completion) require different IT tools. As the pre-project phase relates to the question whether there should be a decision regarding the acceptance of an idea for a project, it is expected that simulation, both visual and numerical, would be a major feature of this stage. Virtual reality, 3D and economic appraisal tools are therefore the most appropriate. The key theme of visualisation is a main feature of this stage. In the pre-construction stage, the client needs are developed into a design solution. This requires the use of AI (to capture knowledge), CAD and VR tools. The key themes of visualisation and intelligence are dominant at this stage. The final stages of construction and operation involve the use of legacy applications such as planning, estimating, etc. The key themes of integration and communication are mainly associated with this stage. It has to be said, however, that most of the IT tools and themes are generic and can be used and associated with any of the phases or stages of the process map (Aouad *et al.*, 1998a).

The IT priority topics list (Figure 17.2) and the IT map (Figure 17.3) have helped to identify five types of technologies which will help the process. These technologies are:

- established and used technologies such as office automation tools, computer-aided planning, estimating, purchasing, etc.;

Figure 17.4 The IT and process intersection

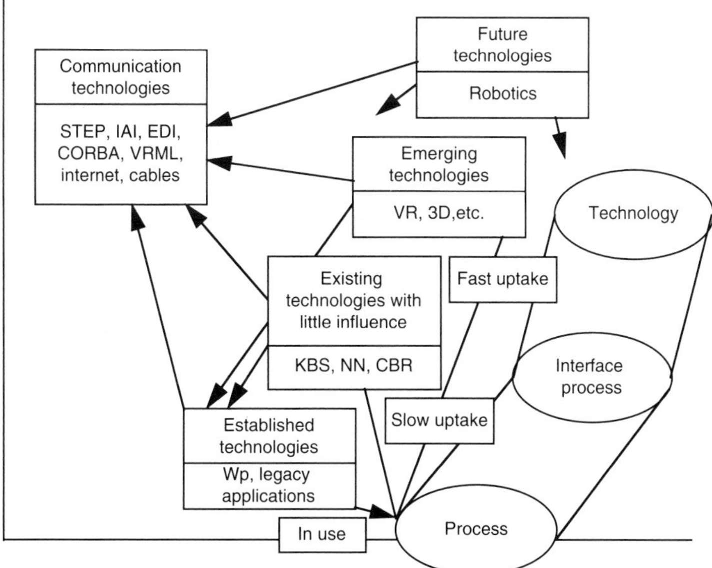

- existing technologies with little current influence such as knowledge-based systems, neural networks, case based technologies, etc.;
- emerging technologies still considered as research tools such as virtual reality, 3D modelling, CASE tools, etc.;
- future technologies such as robotics, automated tools, etc.; and
- communication and standards technologies such as STEP (standard for the exchange of product data), IAI (International Alliance for Interoperability), EDI (electronic data interchange), CORBA (Common Object Request Broker Architecture), VRML (virtual reality modelling language), and Internet communication protocols.

These technologies when integrated will provide the right mechanism for a technology push of the process. However, it has to be said that the emerging and existing technologies will need to have a wider acceptance by the industry before a major impact on the process is realised. This will ultimately result in an improved interface process which will take advantage of the technology and the new ways of performing businesses (Aouad *et al.*, 1997).

Case studies

This section presents the use of IT within two manufacturing and construction projects within the context of the IT map. Both firms have developed process maps. Company A, which is a manufacturing firm, has made use of most of the technologies identified within the IT map. In addition, it has found that the inter-site communications and software applications were facilitated by the provision of an 'all-in-one' computer network. Considering the size of the project team in this particular case study, and the complexity and size of the work that needed to be carried out, it was decided from the beginning of the project that network facilities would need to be utilised. A 'shared' area in the network was allocated to include main documents and information that will be needed by a number of members from the project team. Also, to enable fast and accurate communications, the e-mail addresses of all members were distributed to the project team. Video conferencing facilities were also used to eliminate long travelling periods (more than 4 hours between two sites), for short meetings.

Company B, which is a construction firm, has also been using most of the information technologies identified within the IT map. In particular 3D modelling tools have been used to produce a graphical representation of the buildings involved in the case study. The use of 3D models was mainly for marketing/fund gathering. The electronic and disk-based transfer of information was also adopted as standard practice by some of the participants of the case study project. The benefits associated with the use of IT within this particular case study have been highlighted as follows:

- the archiving and retrieval of project information;
- rapid communications;
- effective coordination;
- the visualisation of client's requirements; and
- the visualisation of structural and spatial requirements.

The expense of the latest technology appears to be the main barrier to the full exploitation of these benefits.

17.4 Summary

This chapter has introduced an IT map (Figure 17.3) which can be looked at as a support tool for a generic design and construction process protocol. This IT map builds upon the results of a recent survey conducted in the UK regarding construction IT priority topics. It is concluded that an IT map should support the key themes of visualisation, intelligence, communication and integration which are vital for a process to survive. The generic and specific IT tools have also been discussed.

The work presented here is still at an early stage. This map should however be used as a stimulator of ideas regarding developing IT solutions for re-engineered design and construction processes. The IT map will serve as a catalyst for the process and should not be looked at as the main driver. The process is driven by needs and requirements and the technology can help in meeting these by providing the right mechanism for visualising, managing, communicating and integrating information which will result in a more consistent and improved process.

18 Capability and maturity in process management

Marjan Sarshar, Alan Hutchinson and Ghassan Aouad

"IT can enable the company to achieve its business objectives."
– American consulting practice director

Much of this book has been written in a general way for how all construction companies can improve their business strategy management by innovating their processes through IT. All companies are different in the extent to which this is achievable from their current position. All companies evolve through different stages of capability maturity. An initiative for standardised process improvement for construction enterprises (SPICE) is applying the capability maturity model from the IT development world to construction process innovation. It demonstrates how ready different companies are to introduce strategic IT systems. The characteristics of mature and immature organisations are very different. Process maturity is a step-by-step migration path through levels of capability.

- Immature construction organisations are reactionary fire-fighters that survive through heroes.
- Mature construction organisations have organisation-wide abilities, good communication, and planned and consistent processes.
- Levels of process maturity range from chaotic, planned and tracked, well defined, supply-chain aligned, qualitatively controlled through to continuously improving.

"For information to be effective it needs to be circulated within the company. The structure of the organisation needs to be flexible, however, if there are more than three reporting levels information stops flowing."
– Director of financial information, French contractor

18.1 Introduction

This final chapter illustrates how IT, and other enablers, need to fit within a business context of process improvement. It recognises that the context in which IT is introduced to strategically exploit process improvement differs considerably between businesses who have reached varying degrees of maturity in their process thinking. For decades, businesses within the construction sector have been applying new methods and technologies to improve productivity and achieve quality gains. Nevertheless, Latham's and Egan's targets for quality and productivity achievements have yet to be achieved. One fundamental limitation to advancement is the challenge of improving the management of construction processes. The synergistic benefits of isolated improvements in methods and technologies cannot be realised in a coordinated and repeatable manner in an undisciplined, chaotic process. Construction projects frequently lack the optimised infrastructure and the support necessary to fulfil Latham's and Egan's challenges. There is a need to develop a discipline for continuous process improvement within the construction industry.

As we have seen in earlier chapters, a process is a sequence of steps performed for a given purpose. The process integrates people, tools and procedures together, as shown in Figure 18.1. Construction processes are what people do, using procedures, methods, tools and equipment to transform materials into a product (e.g. building) that is of value to the client. There is increasing evidence from other industries (Imai, 1986; Paulk, 1993) that continuous process improvement is based on many small, evolu-

Figure 18.1 Components of an integrated process

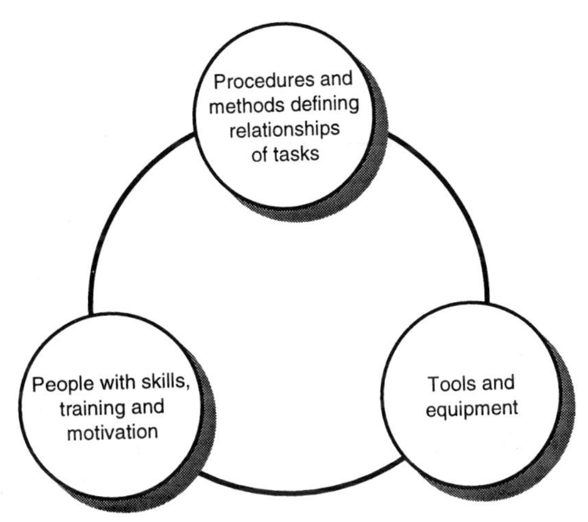

tionary steps, rather than revolutionary measures. Standardised Process Improvement for Construction Enterprises (SPICE) is a new initiative, which attempts to develop an evolutionary step-wise process improvement framework for the construction industry. The SPICE framework will enable construction organisations to improve their processes through various levels of maturity.

18.2 The SPICE initiative

SPICE is based on the experiences gained in the IT sector for step-wise process improvement. SPICE specifically refers to the Capability Maturity Model (CMM) (Paulk, 1993; Saiedian, 1995), which was developed for, and is used by, the US Department of Defense (DoD) who are a major software purchaser. The software industry had contended with poor-quality software, missed schedules, and high costs. In 1991, the Software Engineering Institute (SEI) at Carnegie Mellon University produced the Capability Maturity Model (CMM) for the US Department of Defense (Paulk, 1993). The CMM serves as a framework to continuously measure, evolve and improve processes. CMM has rapidly gained acceptance amongst the IT, telecommunications and engineering based companies in recent years. Successful implementers of CMM have reported high productivity results. Hughes Aircraft (Saiedian, 1995) reported that it spent US$400 000 in a two-year improvement programme. Hughes calculated that its initial return on this investment amounted to US$2 million annually, a 5:1 ROI. Raytheon's numbers are even more remarkable. Investing almost US$1 million annually on improvements, Raytheon (Saiedian, 1995) achieved a 7.7:1 ROI and 2:1 productivity gains. Herlab's (Herlab, 1994) analysis showed that an average of 35% productivity improvements, and an average of 39% post-delivery defect reduction were achieved in companies implementing CMM. These figures show some promise to support Latham's requirements of 30% cost reduction and zero defects by the year 2000 and Egan's calls for 10% annual cost reduction thereafter.

SPICE is developing a similar framework for the construction sector. The CMM framework, as it stands, addresses the needs of information systems development projects. SPICE uses many of the basic concepts of process capability and maturity from CMM. However, SPICE concentrates on construction processes and addresses some construction-specific issues, such as the virtual company nature of the projects, dispersed project teams, and the low ROI in the construction industry, which requires tight business justification of new initiatives. The initial scope of SPICE is medium to large sized design and construction companies. This can then be tailored towards the needs of smaller organisations.

SPICE addresses the processes related to design, construction and maintenance activities within construction companies. It does not concentrate on all the processes related to a company, such as finance or marketing. This is depicted in Figure 18.2. However, experience with CMM demonstrates that through creating an emphasis on core product development processes, organisations gain a process improvement culture. This benefits the whole company, in the long term.

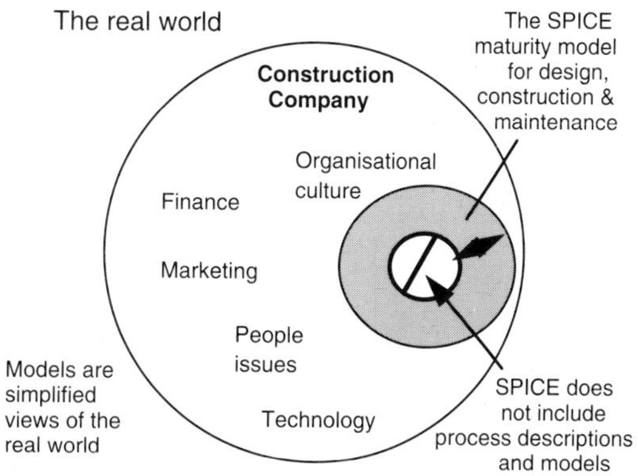

Figure 18.2 The scope of SPICE research

18.3 Immature versus mature construction organisations

Setting sensible goals for process improvement requires an understanding of the difference between immature and mature construction organisations.

In an immature organisation, construction processes are generally improvised by practitioners and project managers during the course of the project. Even if a construction process has been specified, it is not rigorously followed or enforced. The immature organisation is reactionary, and managers are usually focused on fire fighting. In an immature organisation, there is no objective basis for judging product quality or for solving product or process problems. The product quality assurance is often curtailed or eliminated when projects fall behind schedule.

In an immature organisation, product quality is difficult to predict. Activities intended to enhance quality such as reviews are often curtailed. Many of the quality assurance activities are left until the snagging stage at the end of the project. At this point the problems can be too costly to rectify and lead to conflict of interest among the project team.

Even in undisciplined and immature organisations, sometimes individual projects produce excellent results. When such projects succeed, it is generally through the heroic efforts of a dedicated team, rather than through repeating the systematic and proven methods of a mature organisation.

On the other hand, a mature construction organisation possesses an organisation-wide ability for managing design, construction and maintenance activities. The processes are accurately communicated to both existing staff and new employees, and activities are carried out according to the planned processes. The processes mandated are fit for use and consistent with the way the work gets done. Roles and responsibilities within the defined processes are clear throughout the project and across the organisation. In mature organisations, managers monitor the quality of the products and client satisfaction. There is an objective, quantitative basis for judging product quality and analysing problems with the product and process, and a reflective element to the organisational culture. In general, a disciplined process is consistently followed because all of the participants understand the value of doing so, and the necessary infrastructure exists to support the process.

18.4 Process capability and process maturity

Process capability describes the possible range of expected results that can be achieved by following a construction process. The process capability of an organisation provides one means of predicting the most likely outcomes to be expected from the next project, in terms of cost, time and quality. On the other hand, process performance represents the actual results achieved by following a process. It therefore provides historic data on the project.

Since process capability focuses on results expected it can provide predictability for the project. This is an important feature for the clients as well as for the construction organisations. New categories of construction projects often lead to many new surprises and challenges. Process capability of an organisation can provide predictability in the expected outcome of these projects.

Process maturity is the extent to which a specific process is explicitly defined, managed, measured, controlled and effective. Maturity implies a potential for growth in capability and indicates the richness of an organisation's processes and the consistency with which they are applied in projects throughout the organisation. The construction processes are well understood throughout a mature organisation, usually through documentation and training. The processes are continually being monitored and improved by their users. The capability of a mature construction process is known, hence the expected results from the process are predictable. Construction process maturity implies that the productivity and quality of the products can be improved over time through consistent and disciplined focus on process improvement.

As a construction organisation gains process maturity, it institutionalises its construction processes via policies, standards and organisational structures. Institutionalisation entails building an infrastructure and a corporate culture that supports the methods, practices and procedures of the business so that they endure after those who originally defined them have gone.

Based on the context within which a project is conducted, the actual performance of the project may not reflect the full process capability of the organisation, i.e. the capability of the project can become constrained by its environment. Examples of external constraints, which often influence process capability, are economic recessions, new supply chain relationships, or acquisitions and mergers.

18.5 Overview of the framework

Continuous process improvement is based on many small, evolutionary steps (Imai, 1986). The SPICE framework organises these evolutionary steps into maturity levels that lay successive foundations for continuous process improvement. These maturity levels define a scale for measuring the maturity of a construction organisation's processes, and evaluating its process capability. They provide guidelines for prioritising process improvement efforts.

A maturity level is a well-defined evolutionary plateau toward achieving a mature process. Each maturity level provides a layer in the foundation for continuous process improvement. Each level comprises a set of process goals that, when satisfied, stabilise an important component in the 'construction' process. Achieving each level of the maturity framework establishes a different component in the construction process, resulting in an increase in the process capability of the organisation (Paulk, 1993).

The draft of the SPICE maturity levels is depicted in Figure 18.3. This framework is still under development and Figure 18.3 will be refined and improved throughout the research. The maturity framework is split into six levels. At each level some key processes are identified. These key processes are still under development and are not listed in this chapter. An assessment tool accompanies the framework in Figure 18.3. An organisation can only be at one level of maturity at any stage. Organisations conduct an assessment to establish what level of maturity they are at. Judging by the experience in the software industry, most companies are initially at Level 1. They then need to focus on and implement all the key attributes of the next level, i.e. Level 2. The characteristics of each level of the maturity framework are described in the next section.

Figure 18.3 The draft of the SPICE maturity levels

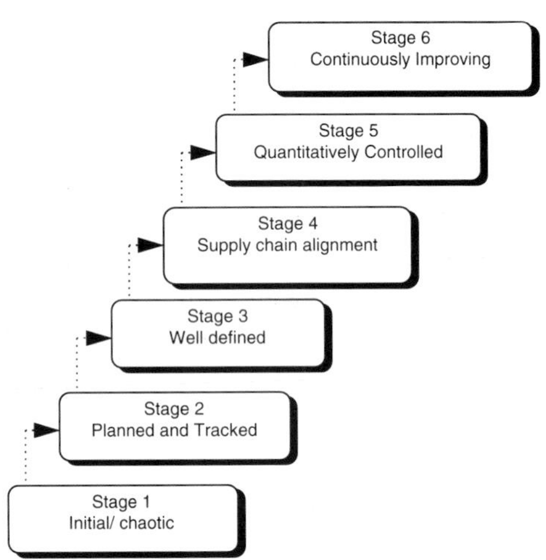

18.5.1 Level 1: Chaotic

At this level project visibility and predictability are poor. Good project practices are local and cannot be repeated across the company in an institutionalised fashion. Ineffective planning and coordination undermine good engineering practices. Organisations make commitments that staff or the supply chain cannot meet. This results in a series of crises.

During a crisis, people on projects typically abandon planned procedures. Time and costs are often under tight control in construction. Hence the crisis often leads to compromises on quality.

Success depends entirely on having an exceptional manager and a competent team. When these managers leave, their stabilising influences leave with them.

The construction process capability of Level 1 organisations is unpredictable because the process is constantly changed or modified as the work progresses. Performance depends on the capabilities of the individuals, rather than that of the organisation.

18.5.2 Level 2: Planned and tracked

At this level there is a degree of project predictability. Policies and procedures for managing the major project-based processes are established. A major objective of Level 2 is to focus on effective management processes within construction projects. This allows organisations to repeat successful practices on earlier projects. An effective process can be characterised as practised, documented, enforced, trained, evaluated and able to improve.

At Level 2 organisations make realistic commitments, based on the results obtained from previous projects and on the requirements of the current project. Managers track quality and functionality as well as time and costs. Problems in meeting the commitments are identified as they arise. The integrity of the project requirements is maintained throughout the project. Standards are defined and organisations ensure that they are faithfully followed. Projects work with subcontractors to establish strong relationships.

18.5.3 Level 3: Well defined

At Level 3 both management and engineering activities are documented, standardised and integrated into the organisation. These standard processes are referenced throughout the organisation. All projects use approved, tailored versions of the organisation's standard processes, which accounts for their unique characteristics.

A well-defined process includes standard descriptions and models for performing the work, verification mechanisms (such as peer reviews) and completion criteria. Because the process is well defined, management has good insight into progress. Quality and functionality of all projects are well tracked.

Up to this stage process improvement efforts are still at an organisational level. They do not address virtual company issues.

18.5.4 Level 4: Supply chain alignment

This level is specific to the construction industry. This level advocates that all the organisations along the supply chain should individually improve their processes up to Level 3. They can now direct their efforts jointly to align the virtual company processes. This level assumes a degree of stability among the supply chain. It will be more successful among companies who are involved in long-term relationships such as partnering relationships or PFI contracts.

The key processes at this level require further research as well as empirical data, before they are finalised.

18.5.5 Level 5: Qualitatively controlled

At this level organisations have the capability to set quality goals for:

- the product;
- the process; and
- the supply chain relationships.

Productivity and quality are measured for important construction process activities across all projects as part of an organisational measurement programme. This forms an objective basis for measuring the product, the process, the degree of customer satisfaction, and the level of harmony across the supply chain.

Projects come under control through narrowing the variations in their process performance to fall within acceptable quantitative boundaries. Meaningful variations can be distinguished from random variations. The risks involved in moving up the learning curve, either due to undertaking new categories of projects or engaging in new procurement and supplier chain arrangements, can be managed.

18.5.6 Level 6: Continuously improving

At this stage the entire supply chain is focused on continuous process improvement. The organisations have the means to identify weaknesses and strengthen the processes proactively, in a collaborative manner. Data on the effectiveness of the processes are used to perform cost-benefit analysis of new technologies and proposed changes to the organisation's processes. Innovations that exploit the best business management practices are identified and transferred throughout the organisations.

Project teams across the supply chain analyse defects to determine their causes. Construction processes are evaluated to prevent known types of defects from recurring, and lessons learned are communicated to other projects.

18.6 SPICE vs CMM

SPICE builds upon the basic concepts of process maturity, which were introduced by CMM. However, SPICE specifically addresses construction processes. The IT industry's processes are different to those of construction. The two industries face different challenges, different cultures and different contractual and supply chain arrangements. Tailoring SPICE to construction requirements is not a trivial task and requires significant input from the construction industry and its representatives.

The draft model depicted in Figure 18.3 already reveals differences with the CMM model. This model attempts to address the supply chain complexity of the construction industry in a step-by-step fashion. Specifically, SPICE has six levels, rather than the five levels of CMM. Level 4, the 'supply chain alignment' level in the SPICE framework, is a new level devised specifically for the needs of the construction industry.

18.7 Common process capability features

The SPICE framework is not prescriptive. It does not tell an organisation how to improve. SPICE describes the major process characteristics of an organisation at each maturity level, without prescribing the means for getting there. The intention is that it does not unduly constrain how the construction processes are implemented by an organisation. It simply describes what the essential process attributes of an organisation would normally be expected to be.

The capability architecture for SPICE is shown in Figure 18.4. This model separates construction-specific processes from capability-related characteristics. The key processes identify the main construction processes to be addressed. As the model develops, the list of these processes will accompany the levels in Figure 18.3.

Each of these key processes requires a disciplined management

Figure 18.4 The SPICE capability architecture

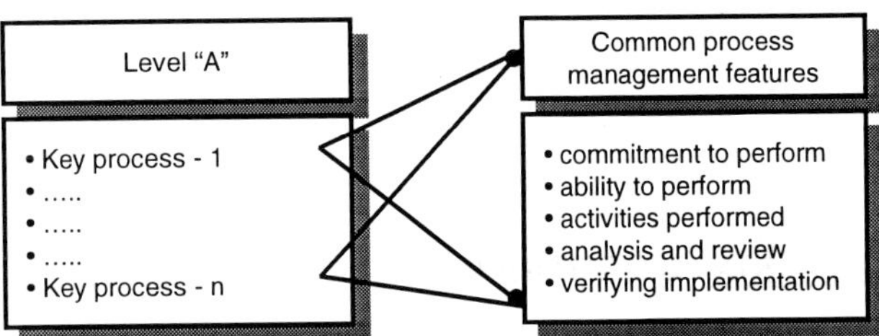

focus in order to be realised. The realisation of these management activities is referred to as the process capability. A process is capable if it possesses a number of features. These features are termed the common process capability features.

The five 'common process capability features' which need to be demonstrated by each key process are listed below.

(1) **Commitment to perform** – This criterion ensures that the organisation takes action to ensure that the process is established and will endure. It typically involves establishing organisation policies. Some processes also require organisational sponsors or leaders. Commitment to perform ensures that leadership positions are created and filled and the relevant organisational policy statements exist.

(2) **Ability to perform** – This describes the preconditions that must exist to implement the process competently. It normally involves adequate resourcing, appropriate organisational structure and training.

(3) **Activities performed** – This describes the activities, roles and procedures necessary to implement processes. It typically involves establishing plans and procedures, performing the work, tracking it, and taking corrective action as necessary.

(4) **Analysis and evaluation** – This describes the basic evaluation practices that are necessary to determine the status of a process. These evaluations are used to control and improve the processes.

(5) **Verifying implementation** – This verifies that the activities are performed in compliance with the process that has been established. It includes reviews by management as well as the quality assurance group.

The SPICE assessment mechanism ensures that each key process has reached capability, by testing it against the above 'common process management features'. For example, a process which is well designed and has a policy statement to support it but is inadequately resourced, will fail the process capability test. This is because the process cannot satisfy the 'ability to perform' requirements.

18.8 Summary

SPICE is an initiative which attempts to develop an evolutionary business process improvement framework for the construction industry. A process assessment tool will accompany this framework. Businesses can use the assessment tool to identify the maturity of their processes. They can then refer to the SPICE framework to establish what their process improvement priorities are likely to be and where to focus their efforts.

The model is not prescriptive. It does not provide any guidelines on how to improve the processes. Instead it provides a set of common process capability features, which all processes need in order to achieve capabilities. These common management features are:

- commitment to perform;
- ability to perform;
- activities performed;
- analysis and evaluation; and
- verifying implementation.

SPICE is based on the Capability Maturity Model (CMM), which is used widely in the IT industry. However, SPICE is construction specific and addresses construction industry issues. The main characteristics of the construction industry which differentiate it from the IT industry are perceived to be:

- the virtual company nature of the projects;
- the dispersity of project teams; and
- the low ROI in the construction industry, which requires tight business justification of new initiatives.

This chapter presents the draft of the SPICE maturity levels. Each of these levels will have a list of 'key processes' associated with it. This list is still under development, and is not presented in this chapter. These key processes will provide guidelines on where to concentrate the process improvement efforts. The importance of this model to this book as a whole is that all companies differ considerably in how well they are able to undertake effective processes through IT, to improve their strategic position.

The first contribution of this chapter is thus to show construction companies how to improve their processes in a systematic way. Beyond this, the chapter is of crucial importance in presenting a methodology for the construction industry to achieve the process redesign and process improvement targets that are called for so repeatedly in the earlier chapters of this book.

Yet the main message of this book is that IT can substantially enable process improvements in construction. This is where the SPICE framework holds the most promise to us. It effectively defines the business process objectives for IT implementation into the future. Our uses of IT in construction over the next ten years will be of enduring value and benefit to construction businesses, and of strategic importance, if they allow all companies to progress in their levels of process maturity. We then need to develop synergistic IT, process and people maturity models in parallel. That is the ultimate challenge for the future with which we should finish this book.

References and bibliography

Abdel-Razek, R.H. and McCaffer, R. (1987) A change in the UK construction industry structure: Implications for tendering. *Construction Management and Economics*, **5**, 227–242.

Andrews, J. (1984) Construction project management in joint ventures in developing countries. *Unibeam*, **XV**, 43–47, Singapore.

Andrews, J. (1990) Building strategy – how to win, structure and manage projects, Paper presented at the First ASEAN International Symposium on Construction and Development, 7–8 May, Kuala Lumpur, Malaysia.

Ansoff, H.I. (1991) Critique of Henry Mintzberg's 'The Design School: Reconsidering the basic premises of strategic management'. *Strategic Management Journal*, **12**, 449–461.

Aouad, G., Alshawi, M. and Bee, S. (1997) Priority topics for construction IT research. *The International Journal of Construction IT*, Winter.

Aouad, G., Hinks, J., Cooper, R., Sheath, D., Kaglioglou, M. and Sexton, M. (1998a) An IT map for a generic design and construction process protocol. *International Journal of Construction Procurement.*

Aouad, G., Kaglioglou, M., Cooper, R., Sexton, M. and Hinks, J. (1998b) Technology management of IT in construction: A driver or an enabler? *Journal of Logistics and Information Management.*

Aouad, G., Cooper, R., Kaglioglou, M., Hinks, M. and Sexton, M. (1998c) A synchronised process/IT model to support the co-maturation of processes and IT in the construction sector. *Proceedings of CIB W78 Conference on IT for Construction*, Sweden.

Armstrong, J.S. (1985) *Long-Range Forecasting: From Crystal Ball to Computer*, 2nd edn., Wiley-Interscience, New York.

Atkin, B.L., Clark, A.M. and Gravett, J.V.J. (1997) *Benchmarking Best Practice IT in Facilities Management*, Construct IT Centre of Excellence, University of Salford.

Avison, D.E. and Fitzgerald, G. (1995) *Information Systems Development: Methodologies, Techniques and Tools.* 2nd edn., McGraw Hill International (UK) Limited.

Avison, D.E. and Shah, H. (1997) *The Information Systems Development Life Cycle: a First Course in Information Systems.* McGraw Hill International (UK) Limited.

ASCE Aerospace Division (1988) Civil engineers in space. *Journal of Professional Issues in Engineering*, **114**, No. 3, 348–353.

Benjamin, R.I., Rockart, J.F., Scott Morton, M.S. and Wyman, J. (1984)

Information technology: A strategic opportunity. *Sloan Management Review*, Spring, 3–10.

Bennett, J. (1992) Book Review of Strategic Management in Construction, *Construction Management and Economics*, **10**, No. 6, 533–534.

Betts, M. (1992) How Strategic is our use of Information Technology in the Construction Sector? *International Journal of Information Technology in Construction*, **1** No. 1.

Betts, M. and Ofori, G. (1992) Strategic Planning for Competitive Advantage in Construction. *Construction Management and Economics*, **10**, No. 6, 511–532.

Betts, M. and Ofori, G. (1994) Strategic planning for competitive advantage in construction: The institutions. *Construction Management and Economics*, **12**, 203–217.

Betts, M., Mathur, K. and Ofori, G. (1989) *Information Technology and the Construction Industry of Singapore: A Framework for A Communications Network*. School of Building and Estate Management, National University of Singapore.

Betts, M., Clark, A., Grilo, A. and Miozzo, M. (1995) Occasional Paper Number 1: A process-based study of an IT research work plan. Construct IT Centre of Excellence, University of Salford.

Bjork, B. (1985) *Computers in Construction; Research, Development and Standardisation Work in the Nordic Countries*. Technical Research Centre of Finland.

Bjork, B. (1987) *The Integrated Use of Computers in Construction – The Finnish Experience*. Nordic Building Research (NBS).

Bjornsson, H. and Lundegard, R. (1992) Corporate Competitiveness and Information Technology. *European Management Journal*, **10**, No. 3, 341–347.

Bjornsson, H. and Lundegard, R. (1993) Strategic Use of IT in some European Construction Firms. In: *Management of IT for Construction*, World Scientific Publishers, Singapore.

Bonn, I. and Christodoulou, C. (1996) From Strategic Planning to Strategic Management, *Long Range Planning*, August, **29**, No. 4, 543–551.

Boston Consulting Group (1994) *Reengineering and Beyond*, BCG.

Bradshaw, D. (1990) Building Blocks of Efficiency, *Financial Times*, p. 14 August 30.

Brandon, P.S., (1990) The Development of an Expert System for the Strategic Planning of Construction Projects, *Construction Management and Economics*, **8**, No. 3, 285–300.

Brandon, P. and Betts, M. (1995) *Integrated Construction Information*. E & FN Spon, London.

Breuer, J. and Fischer, M. (1994) Managerial aspects of information technology strategies for AEC firms. *Journal of Management in Engineering*, July/August, **10**, 4.

British Airports Authority (1994) *The Project Process*. BAA plc, London.

British Property Federation (1983) *Manual of the BPF System*. The British Property Federation System for Building Design and Construction (BPF), London.

Brochner, J. (1990) Impacts of Information Technology on the Structure of Construction. *Construction Management and Economics*, **8**, 205–218.

Bruns, W.J. and McFarlan, F.W. (1987) Information Technology puts power in control systems. *Harvard Business Review*, Sept–Oct, **65**, No. 5, 89–94.

Buckingham, R.A., Hirschheim, R.A., Land, F.F. and Tully, C.J. (eds) (1987) *Information Systems Education: Recommendations and Implementation*. Cambridge University Press, Cambridge.

Bullen, C.V. and Rockhart, J.F. (1984) A Primer on Critical Success Factors, *Information Systems Working Paper (No. 69)*. Sloan School of Management, MIT.

Burberry, P. (1991) Saving energy: What matters now? *Architect's Journal*, 13 February, 55–59.

Businessweek (1983) Business is turning data into a potent strategic weapon, *Businessweek*, August 22nd, 50–52.

Cash, J.I. and McFarlan, F.W. (1990) *Competing Through Information Technology*. Harvard Business School Video Series, Harvard, Boston.

Cash, J.I., McFarlan, F.W. and McKenney, J.L. (1983) *Corporate Information Systems Management: Text and Cases.* Richard Irwin, Illinois.

Centre for Strategic Studies in Construction (1988) *Building Britain 2001*. Reading.

Channon, D.F. (1978) *The Service Industries*. Macmillan, London.

Chartered Institute of Building (1989) *What Are You Doing About the Environment?* CIOB, Ascot, UK.

Chartered Institute of Building (1995) *Report of a Study Mission to Japan*. CIOB, Ascot, UK.

Chatzoglon, P.D. and Macaulay, L.A. (1996) Requirements Capture and IS Methodologies. *Information Systems Journal*.

Checkland, P. (1981) *Systems Thinking, Systems Practice*. Wiley, Chichester.

Checkland, P. and Scholes, J. (1990) *Soft Systems Methodology in Action*. Wiley, Chichester.

Checkland, P., Clarke, S. and Poulter, J.S. (1996) The Use of Soft Systems Methodology for Developing HISS and IM&T Strategies in NHS Trusts. *Healthcare Computing*.

Chow, K.F. (1989) The Impact of Information Technology on the Construction Industry. *Proceedings of the First IES Information Technology Conference*, 25–27 May, Singapore.

Chow, K.F. (1990) *The Construction Agenda: Development of the construction industry in Singapore*. Construction Industry Development Board, Singapore.

Clark, A., Young, H.K., Grilo, A., Betts, M. and Ibbs, W. (1995) Contemporary Strategic Planning Tools and Applications for Construction Managers, *First International Conference on Construction Project Management*, January, Singapore.

Codington, S. and Wilson, T.D. (1994) Information Systems Strategies in the UK Insurance Industry, *International Journal of Information Management*, **14**, 188–203.

Collins, J. and Porras, J. (1991) Organisational Vision and Visionary Organisations, Research Paper 1159, Graduate School of Business, Stanford University.

Collis, D. and Montgomery, C. (1995) Competing on resources: strategy in the 1990s, *Harvard Business Review*, July, 73, No. 4, 118.

Construct IT Centre of Excellence (1995a) *Best Practice Benchmarking Report, Supplier Management*. Construct IT Centre of Excellence, Salford, UK.

Construct IT Centre of Excellence (1995b) *A Process-Based Analysis of an IT Research Work Plan*, Occasional Paper 1. Construct IT Centre of Excellence, Salford, UK.

Construct IT Centre of Excellence (1996a) *Construction Site Processes Benchmarking Report*. Construct IT Centre of Excellence, Salford, UK.

Construct IT Centre of Excellence (1996b) *Briefing and Design Benchmarking Report*. Construct IT Centre of Excellence, Salford, UK.

Construct IT Centre of Excellence (1997a) *The Armathwaite Initiative: Global Construction IT Futures International Meeting*. Construct IT Centre of Excellence, Salford, UK.

Construct IT Centre of Excellence (1997b) *A Health Check of the Strategic Exploitation of IT*. Construct IT Centre of Excellence, Salford, UK.

Construct IT Centre of Excellence (1997c) *Facilities Management Benchmarking Report*. Construct IT Centre of Excellence, Salford, UK.

Construct IT Centre of Excellence (1998a) *Supplier Management Update Benchmarking Report*. Construct IT Centre of Excellence, Salford, UK.

Construct IT Centre of Excellence (1998b) *Cost and Change Management Benchmarking Report*. Construct IT Centre of Excellence, Salford, UK.

Construction Industry Development Board (1989) *Report on the Cost Competitiveness of the Construction Industry of Singapore*. Singapore.

Construction Industry Vision Study Group (1986) *A Vision of the Construction Industry in the 21st Century*. Ministry of Construction, Tokyo.

Construction IT Forum (1995) *IT 2005*. CIT Forum, Cambridge.

Coombs, R. (1996) Core competencies and the strategic management of R&D, *R&D Management*, October, **26**, No. 4, 345–355.

Cooper, R., Kaglioglou, M., Aouad, G., Hinks, J., Sexton, M. and Sheath, D. (1998) Development of a Generic Design and Construction Process. *Proceedings of the European Conference on Product Data Technology*, pp 205–214, Building Research Establishment.

Cooper, W.W. (1992) On Porter's Competitive Advantage of Nations. *Omega*, **20**, No. 2, 137–138.

Coordinating Committee for Project Information (1987) *Coordinated Project Information for Building Works, a guide with examples*. CCPI, Royal Institution of Chartered Surveyors, London.

Craig, J.C. and Grant, R.M. (1995) *Strategic Management*. Kogan Page.

CSC Index (1994) *The State of Reengineering Report: Executive Summary*. CSC Index.

Daniels, C. (1991) *The Management Challenge of Information Technology*. The Economist Intelligence Unit Management Guides, London.

Davenport, T.H. (1993) *Process Innovation: Reengineering Work Through Information Technology*. Harvard Business School Press, Boston.

Davenport, T.H. and Short, J.E. (1990) The New Industrial Engineering: Information Technology and Business Process Redesign. *Sloan Management Review*, **31**, No. 4, 11–27.

Davenport, T.H. and Stoddard, D.B. (1994) Reengineering: Business Change of Mythic Proportions. *MIS Quarterly*, June.

Davids, A. (1992) *Practical Information Engineering: The Management Challenge*. Pitman Publishing, London.

Department of the Environment (1995) *Bridging the Gap – An Information Technology Strategy for the UK Construction Industry*. HMSO, London.

Department of the Environment (1996a) *Bridging the Gap: Implementation Plan 1996–2001*, HMSO, London.

Department of the Environment (1996b) *Bridging the Gap: The Feasibility of a Library of Building Design Objects*, HMSO, London.

Department of the Environment (1996c) *Bridging the Gap: The Feasibility of an Industry Knowledge Base*, HMSO, London.

Department of the Environment (1996d) *Bridging the Gap: The Feasibility of the Integrated Project Database*, HMSO, London.

Department of Trade and Industry (1992) *Best Practice Benchmarking: An Executive Guide*. Department of Trade and Industry.

Dioguardi, G. (1983) Macrofirms: Construction firms for the computer age. *Journal of Construction Engineering and Management*, **109**, No. 1, March, 13–24.

Dixon, N. (1993) *Strategic Planning for Information Technology in Construction Organisations: Preparing for the Information Era*. MSc Dissertation, Department of Surveying, University of Salford.

Earl, M.J. (1989) *Management Strategies for Information Technology*. Prentice Hall, Hemel Hempstead.

Earl, M.J. (1993) Experiences in Strategic Information Systems Planning. *MIS Quarterly*, March.

Eason, K. (1988) *Information Technology and Organisational Change*. Taylor and Francis, London.

Economic Committee (1986) *The Singapore Economy: New Directions*. Ministry of Trade and Industry, Singapore.

Egan, Sir J. (1998) *Rethinking Construction*. Report of the Construction Task Force to John Prescott, Department of Environment, Transport and the Regions, London.

Ehrenberg, D. (1989) Decision support systems for management in central warehouses of building plants. *Engineering Costs and Production Economics*, **15**, May, 181–184.

Eilon, S. (1992a) Editorial: On Competitiveness. *Omega*, **20**, No. 1, i–v.

Eilon, S. (1992b) Editorial: Intrepid Gurus. *Omega*, **20**, No. 4, 411–415.

Engineering News Record (1986) Top International Contractors, July 17.

Fellows, R.F. (1993) Competitive Advantage in Construction: Comment. *Construction Management and Economics*, **11**.

Finch, E., Flanagan, R. and Marsh, L. (1996) Auto-ID application in Construction. *Construction Management and Economics*, **14**, 121–129.

Finkelstein, C. (1989) *An Introduction to Information Engineering: From Strategic Planning to Information Systems*. Addison-Wesley, Sydney.

Flaaten, P.O., McCubbrey, D.J., O'Riordan, P.D. and Burgess, K. (1989) *Foundations of Business Systems*. Arthur Andersen & Co. and Dryden Press, Florida.

Foss, N. (1996) Research in strategy, economics and Michael Porter. *Journal of Management Studies*, January, **33**, 1–24.

Gallagher, J.P. (1988) *Knowledge Systems for Business*. Prentice-Hall, New York.

Galliers, R.D. (1987) Information Technology for Comparative Advantage: Serendipity or Strategic Vision? Keynote address, AUSCAM National Conference, Perth, Australia, 23 October.

Galliers, R.D. (1991) Strategic Information Systems Planning: Myths, Reality and Guidelines for Successful Implementation. *European Journal of Information Systems*, **1**, No. 1.

Garvin, D. (1995) Leveraging processes for strategic advantage. *Harvard Business Review*, September, **73**, No. 5, 76.

Gerstein, M.S. (1988) *The Technology Connection: Strategy and Change in the Information Age*. Addison-Wesley, New York.

Gibb, Sir F. (1990) Pollution and its containment. *Overseas Brief* (Newsletter of the Institution of Civil Engineers), September, pp. 1–2.

Gibson, G.E. and Bell, L.C. (1990) Electronic Data Interchange in Construction. *Journal of Construction Engineering and Management*, **116**, No. 4, December, 727–737.

Gibson, V. (1995) Align people with business strategies. *HR Focus*, **72**, No. 3, 12.

Gillin, P. (1990) Building IS from the ground up. *Computerworld*, August 20, pp. 57–61.

Ginsberg, A. (1997) New age strategic planning: Bridging theory and practice. *Long Range Planning*, **30**, No. 1, 125–128.

Glaser, B. and Strauss, A. (1967) *The Discovery of Grounded Theory*. Aldine, Chicago.

Gonzales, A., Ogunlana, S. and Soegaard, R. (1993) Technology impact grid: a model for strategic IT planning for competitive advantage in construction. In: *Management of IT for Construction*, World Scientific Publishers, Singapore.

Goold, M. (1992) Design Learning and Planning: A Further Observation on the Design School Debate. *Strategic Management Journal*, **13**, 169–170.

de Gues, A. (1997). The Living Company. *Harvard Business Review*, March–April.

Hall, G., Rosenthal, J. and Wade, J. (1993) How to Make Reengineering Really Work. *Harvard Business Review*, November–December, 119–131.

Halpin, D.W. (1990) International competition in construction technology. *Journal of Professional Issues in Engineering*, **116**, No. 4, October, 351–359.

Hammer, M. (1990) Reengineering Work: Don't Automate, Obliterate. *Harvard Business Review*, July–August, **68**, No. 4, 104–122.

Hammer, M. and Champy, J. (1993) *Reengineering the Corporation: A Manifesto for Business Revolution*. Harper Business, New York.

Hammer, M. and Stanton, S.A. (1995) *The Reengineering Revolution*. Harper Collins Publishers, London.

Hampson, K. and Tatum, C. (1997) Technology strategy and competitive performance in bridge construction. *Journal of Construction Engineering and Management*, **123**, No. 2, 153.

Hansen, K. and Tatum, C. (1996) How strategies happen: a decision-making framework. *Journal of Management in Engineering*, **12**, No. 1, 40.

Hartasanchez-Garana, J. (1985) Mexican experience in the use of micro-computers on the construction site. In: *Proceedings of the Second International Conference on Civil and Structural Engineering Computing*, Civil-Comp 85, (ed. B.H.V. Topping), Vol. 1 pp. 53–57.

Hasegawa, F. (1988), *Built by Japan: Competitive strategies of the Japanese construction industry*. John Wiley and Sons, New York.

Hemmett, F.J. (1991) Computer linked information exchange network technology - a practical example. In: *Applications of Information Technology in Construction*, Thomas Telford Ltd, London, pp. 49–63.

Herbleb, J. (1994) Software Process Improvement: State of the Payoff. *American Programmer*, September.

Higgin, G. and Jessop, N. (1975) *Communications in the Building Industry*. Tavistock, London.

Hillebrandt, P.M. (1990), Management of the building firm. *Proceedings, CIB 90, Joint Symposium on Building Economics and Construction Management*, Sydney, March, Vol. 6, pp. 1–10.

Hillebrandt, P.M. and Cannon, J. (1990) *The Modern Construction Firm*. Macmillan, London.

Hinks, J., Aouad, G., Cooper, R., Sheath, D., Kaglioglou, M. and Sexton, M. (1997) IT and The Design and Construction Process: A Conceptual Model of Co-Maturation. *The International Journal of Construction IT*, July.

Hofman, S.S. (1987) Tishman builds on MIS cornerstone. *Information Week*, No. 129, August 10, pp. 27–28.

Howard, H.C. (1991) Project-Specific Knowledge Bases in AEC Industry. *Journal of Computing in Civil Engineering*, **5**, No. 1, January, 25–41.

Imai, M. (1986) *Kaizen: The Key to Japan's Competitive Success*. McGraw-Hill, New York.

International Council for Building Research and Documentation (1989) *Trends in Building Construction Techniques Worldwide*. Rotterdam.

Ives, B. and Learmonth, G.P. (1984) The information system as a competitive weapon. *Communications of the ACM*, pp. 1193–1201.

Jennings, M. and Betts, M. (1996) Competitive strategy for quantity surveying practices: the importance of IT. *Engineering Construction and Architectural Management*, **3**, 163–186.

Johansson, H.J., McHugh, P., Pendlebury, A.J. and Wheeler, W.A. (1993)

Business Process Reengineering: Breakpoint Strategies for Market Dominance. John Wiley & Sons, Chichester.

Kagliogiou, M., Cooper, R., Aouad, G., Hinks, J., Sexton, M. and Sheath, D. (1998a) Cross-Industry Learning: The development of a Generic Design and Construction Process Based on Stage/Gate New Product Development Processes Found in the Manufacturing Industry. *Proceedings of the Engineering Design Conference,* Brunel University, June.

Kagliogiou, M., Aouad, G., Cooper, R. and Hinks, J. (1998b) The Process Protocol: Process and IT Modelling for the UK Construction Industry. *Proceedings of the Second European Conference on Product and Process Modelling in the Building Industry,* October, Building Research Establishment, Watford.

Kagliogiou, M., Cooper, R. and Aouad, G. (1999) The Process Protocol: Improving the Front End of the Design and Construction Process for the UK Industry. *Harmony & Profit, CIB Working Commission W92, Procurement Systems Seminar,* Chiang Mai, Thailand, January 25–28.

Kaplan, R. and Norton, D. (1996) Using the balanced scorecard as a strategic management system. *Harvard Business Review,* January, **74**, No. 1, 75.

Kearney, A.T. (1984) The Barriers and the Opportunities of Information Technology – a Management Perspective. A management consultant report prepared for the Department of Trade and Industry, London.

King, W.R. (1988) How Effective is your Information Systems Planning? *Long Range Planning,* **21**, 5.

Koskela, L. (1985) *Construction Industry: Towards the Information Society; The Japanese Example.* FACE (International Federation of Associations of Computer Users in Engineering, Architecture and Related Fields), Report No. 7, November, Finland.

Koskela, L. (1992) *Application of the New Production Philosophy to Construction.* CIFE Technical Report No. 72, Center for Integrated Facility Engineering, Stanford University.

KPMG and CICA (1993) *Building on IT for Quality.* London.

Krippendorf, K. (1980) *Content Analysis: An Introduction to its Methodology.* Saga Publications, London.

Langford, D. (1992) Book Review of International Construction Project Management General Theory and Practice. *Construction Management and Economics,* **10**, No. 2, 179–181.

Langford, D. and Male, S. (1991) *Strategic Management in Construction.* Gower, Aldershot.

Lansley, P. (1983) A practical approach to auditing organizational flexibility. *Construction Management and Economics,* **1**, 145–156.

Lansley, P. (1987) Corporate strategy and survival in the UK construction industry. *Construction Management and Economics,* 5, pp. 141–155.

Lansley, P., Quince, T. and Lea, E. (1979) *Flexibility and Efficiency in Construction Management.* Ashridge Management Research Unit, Berkhampstead.

Latham, Sir M. (1994) *Constructing the Team.* HMSO, London.

Lea, E., Lansley, P. and Spencer, P. (1974) *Growth and Efficiency in the Building Industry*. Ashridge Management Research Unit, Berkhampstead.

Lederer, A.L. and Sethi, V. (1988) The Implementation of Strategic Information Systems Planning Methodologies. *MIS Quarterly*, September.

Lederer, A.L. and Sethi, V. (1992), Meeting the Challenges of Information Systems Planning. *Long Range Planning*, **25**, 2.

Lee, K.C. (1988) The computerization of the Singapore civil service. *Proceedings of Expert Group Meeting: Integrating Information Systems/Technology in Local/Regional Development Planning*, Singapore, 31 October–4 November.

Leibfried, K.H.J. and McNair, C.J. (1994) *Benchmarking: A Tool for Continuous Improvement*, HarperCollins, London.

Lewis, J.C. and Naim, M.M. (1995) Benchmarking of aftermarket supply chains. *Production Planning and Control*, **6**, No. 3, 258–269.

Lim, S.S. (1990) *Singapore's Opportunities for Competitive Advantage*. Keynote Address to Opening Session of the National IT Application Conference, 1–7 March, Singapore.

Love, P. (1996) Enablers of process re-engineering. *Proceedings of InCIT '96*, pp. 77–83, Sydney, Australia.

Madlin, N. (1987) Kellogg's Incon: Engineered for competition. *Management Review*, **76**, No. 7, July, 60–61.

Maister, D. (1986) The Three Es of Professional Life. *Journal of Management Consulting*, **3**, No. 2, 39–44.

Male, S. and Stocks, R. (1991) *Competitive Advantage in Construction*. Butterworth Heinemann, Oxford.

Maloney, F. (1997) Strategic planning for human resource management in construction. *Journal of Management in Engineering*, May, **13**, No. 3, 49.

Mathur, K., Betts, M. and Tham, K.W. (1993) *The Management of IT for Construction*. World Scientific, Singapore.

McDonagh, N. (1991) Future shock or future shocked. *Chartered Quantity Surveyor*, June, pp. 10–12.

McFarlan, F.W. (1984) Information technology changes the way you compete. *Harvard Business Review*, May–June, 98–104.

McHugh, P., Merli, G. and Wheeler, W.A. (1995) *Beyond Business Process Reengineering: Towards the Holonic Enterprise*. Wiley.

McKenney, J.L. and McFarlan, F.W. (1982) The Information Archipelago – Maps and Bridges. *Harvard Business Review*, September–October, 109–120.

Milliken, J. (1997) Strategic Management: Concepts and Cases. *Long Range Planning*, April, **30**, No. 2, pp. 305–306.

Mintzberg, H. (1991) Learning 1, Planning 0. *Strategic Management Journal*, **12**, pp. 463–466.

Miyatake, Y. (1996) Technology development and sustainable construction, *Journal of Management in Engineering*, July, **12**, No. 4, p. 23.

Moran, S. (1984) Adios, Adobe. *Viewpoint*, **12**, No. 3, May/June, pp. 24–26.

Mulcahy, J.F. (1990) Management of the building firm. *Proceedings, CIB 90,*

Joint Symposium on Building Economics and Construction Management, Sydney, March, Vol. 6, pp. 11–21.

Mulqueen, J.T. (1987) Contractor Links Back Office to Construction Sites. *Data Communications*, **16**, No. 3, March, pp. 123–124.

Nam, C. and Tatum, C. (1997) Leaders and champions for construction innovation. *Construction Management and Economics*, **15**, pp. 259–270.

National Computer Board (1991) *IT2000: Construction and Real Estate Sector Study Report*, Singapore.

New Civil Engineer/New Builder (1995) BAA Supplement. Thomas Telford, London.

Newcombe, R. (1990) The evolution and structure of the construction firm. *Proceedings, CIB 90, Joint Symposium on Building Economics and Construction Management*, Sydney, March, Vol. 6, pp. 358–369.

New South Wales Government (1998a) Information Technology in Construction - making IT happen, Green Paper, April.

New South Wales Government (1998b) Construct New South Wales: Seizing Opportunities to Build a Better Construction Industry, White Paper, July.

O'Brien, M. (1996) A strategy for achieving data integration in construction. *International Journal of Construction IT*, **4**, No. 1, 21–34.

Office of Science and Technology (1995) *Technology Foresight: Report of the Construction Sector Panel*, HMSO, London.

Ofori, G. (1990) *The Construction Industry: Aspects of its economics and management*. Singapore University Press, Singapore.

Ofori, G. (1992) The environment: The fourth construction project objective? *Construction Management and Economics*, **10**, No. 5, 369–395.

Ofori, G. (1994) Strategic Planning in the Singapore Construction Industry, *Construction Management and Economics*, **12**.

Ofori, G., Betts, M. and Mathur, K. (1991) Implementing an Information Technology Framework for the Construction Industry of Singapore. *Australian Institute of Building Papers*, **4**, 133–150.

Ohmae, K. (1990) *The Borderless World*. Harper Business, New York.

Osterman, P. (1991) The Impact of IT on Jobs and Skills. In: *The Corporation of the 1990s: Information Technology and Organisational Transformation* (ed. M.S. Scott-Morton), Oxford University Press.

Parsons, G.L. (1983) Information technology: A new competitive weapon. *Sloan Management Review*, Fall, pp. 3–14.

Paulk, M.C. et al. (1993) Capability Maturity Model for Software, version 1, Software Engineering Institute, Carnegie Mellon University, February.

Paulk, M.C., Weber, C.V., Curtis, B. and Chrissis, M.B. (1995) *The Capability Maturity Model: Guidelines for Improving the Software Process*, Addison-Wesley.

Peat, Marwick and McLintock and the Construction Industry Computing Association (1990) *Building on IT for the 1990s: A Survey of the Information Technology Trends and Needs in the Construction Industry*. Peat Marwick and McLintock, London.

Penrose, E.T. (1966) *The Theory of the Growth of the Firm*. Blackwell, Oxford.

Perkowski, J.C. (1988) Technical trends in the E&C business: The next 10 years. *Journal of Construction Management and Engineering*, **114**, No. 4, pp. 565–576.

Petrozzo, D.P. and Stepper, J.C. (1994) *Successful Reengineering*. Van Nostrand Reinhold.

Peters, A. (1991) Orwell's Robots. *Asia*, **29**, No. F-10, February, pp. 6–15.

Porter, M.E. (1979) How competitive forces shape strategy. *Harvard Business Review*, March–April, pp. 137–146.

Porter, M.E. (1980) *Competitive Strategy*. Free Press, New York.

Porter, M.E. (1985) *Competitive Advantage*. Free Press, New York.

Porter, M.E. (1990) *The Competitive Advantage of Nations*. Free Press, New York.

Porter, M.E. and Millar, V.E. (1985) How information gives you competitive advantage. *Harvard Business Review*, July–August, pp. 149–160.

Prahalad, C.K. and Hamel, G. (1990) The core competence of the corporation. *Harvard Business Review*, May-June, pp. 79–91.

Premkumar, G. and King, W.R. (1991) Assessing Strategic Information Systems Planning. *Long Range Planning*, October.

Pugh, D.S. (1971) *Organisation Theory: Selected Readings*. Penguin Books. London.

Raman, K.S. (1988) Application of IT in small and medium sized enterprises in Singapore. *Proceedings of Expert Group Meeting: Integrating Information Systems/Technology in Local/Regional Development Planning*, Singapore, 31 October–4 November.

Ramsay, W. (1989) Business objectives and strategy. In *The Management of Construction Firms: Aspects of Theory*, (ed. P.M. Hillebrandt and J. Cannon), Macmillan, London.

Rashid, A. (1991) Global Strategies of Construction Firms. Unpublished PhD thesis, University of Reading.

Ray-Jones, A. (1990) The why, what and how of a graphics library. *Proceedings of the National IT Application Conference*, Construction Sector Programme, 1–7 March, Singapore.

Ridgway, M.R. (1998) MSc thesis, Department of Surveying, University of Salford.

Ridgway, M.R., Dunnett, S., Holden, T. and Taylor, R. (1997) Does Your IT Solution Fit Your Business: An Information Strategy Plan for Bramall Construction Ltd. MSc project, Department of Surveying, University of Salford.

Royal Institution of Chartered Surveyors (1979) *UK and US Construction Industries*. RICS Publications, London.

Royal Institution of Chartered Surveyors (1998) *The Challenge of Change*, Report of the QS Think Tank, RICS Publications, London.

Rumelt, R.P., Schendel, D. and Teece, D.J. (1991) Strategic Management and Economics. *Strategic Management Journal*, **12**, 5–29.

Sadler, P., Webb, T. and Lansley, P. (1974) *Management Style and Organisation Structure in the Smaller Enterprise*, Ashridge Management Research Unit, Berkhampstead.

Saiedian, H. and Kuzara, N. (1995) SEI Capability Maturity Model's Impact on Contractors, *IEEE Computer*, Jan.

Sanvido, V. (1992) A top down approach to integrating the building process. *Engineering with Computers*, **5**, 91–103.

Sarshar, M., Powell, J., Aouad, G., Brandon, P., Brown, F., Cooper, G., Ford, S., Kirkham, J. and Young, B. (1993) Object Orientation and Information Engineering: A Hybrid Approach to Information Modelling in Construction. *International Journal of Construction IT*, Summer, 83–96.

Sarshar, M., Aouad, G. and Brandon, P. (1994) Improving Project Organisation Through Strategic Planning of Information. *International Journal of Construction IT*, **2**, No. 3, Autumn, 77–94.

Schwartz, P. (1991) *The Art of the Long View*. Doubleday Currency.

Scott-Morton, M.S. (1991) *The Corporation of the 1990s: Information Technology and Organisational Transformation*. Oxford University Press.

Simister, S. (1993) Computer assisted analysis of interviews for construction management research. *Proceedings of the 9th Annual ARCOM Conference*, (eds R.A. Eastham and R.M. Skitmore), University of Salford.

Soat, J. (1988) How MIS became the business. *Computer and Communications Decisions*, **20**, No. 2, February, 83–85.

Stalk, G., Evans, P. and Shulman, L.E. (1992) Competing on Capabilities: The New Rules of Corporate Strategy. *Harvard Business Review*, March–April, 57–69.

Stowell, F.A. (ed.) (1995) *Information Systems Provision: The Contribution of Soft Systems Methodology*. McGraw-Hill, London.

Takenaka Corporation (1991) Annual Report, Tokyo.

Tapscott, D. (1996) The *Digital Economy: Promise and Peril in the Age of Networked Intelligence*. McGraw-Hill, New York.

Tatum, C.B. (1987) The process of innovation in the construction firm. *Journal of Construction Engineering and Management*, American Society of Civil Engineers, **113**, No. 4, 648–663.

Tatum, C.B. (1988) Technology and competitive advantage in civil engineering. *Journal of Professional Issues in Engineering*, American Society of Civil Engineers, **114**, No. 3, July.

Tatum, C.B. (1990) Integration: Emerging management challenge. *Journal of Management in Engineering*, **6**, No. 1, January, 47–58.

Teicholz, P. and Fischer, M. (1994) Strategy for Computer Integrated Construction Technology. *Journal of Construction Engineering and Management*, **120**, No. 1, pp. 117–131.

Tesch, R. (1990) *Qualitative research: analysis types and software tools*, The Falmer Press, London.

The Economist (1990a) Builders take their partners for Europe, July 14, pp. 63–64.

The Economist (1990b) Porter v Ohmae, August 4, p. 59.

The Economist (1991a) Competing with tomorrow, May 12, pp. 67–68.

The Economist (1991b) Gordon Wu: Building East Asia, October 19, p. 91.

The Economist (1991c) The Vision Thing, November 9, p. 89.

The Economist (1994) Re-engineering Europe, February 26, pp. 81–82.

Tozer, E.E. (1996) *Strategic IS/IT Planning*. Butterworth-Heinemann, Newton, MA.

Turner, B. (1981) Some practical aspects of qualitative data analysis: one way of organising the cognitive process associated with the generation of grounded theory. *Quality and Quantity*, **15**, Elsevier Scientific, Amsterdam.

Turner, B. (1983) The use of grounded theory for the qualitative analysis of organisational behaviour. *Journal of Management Studies*, **20**, No. 3. pp. 333–348.

Venkatranam, N. (1991) *IT-induced Business Reconfiguration in the Corporation of the 1990s: Information Technology and Organisational Transformation*, Oxford University Press.

Vidgen, R., Rose, J., Wood-Harper, T. and Wood, J.R.G. (1994) Business Process Reengineering: The Need for a Methodology to Revision the Organisation. In: *Business Process Reengineering: Information Systems and Opportunities*, Elsevier Science, North-Holland, pp. 603–612.

VTT (1990) Information and Automation Systems in Construction, Research Programme 1988–90, Technical Research Centre of Finland, Espoo.

Ward, J. and Griffiths, P. (1996) *Strategic Planning for Information Systems*. 2nd edn, Wiley, Chichester.

Williams, S. (1989) *Hong Kong Bank*. Jonathan Cape, London.

Wilson, B. (1990) *Systems: Concepts, Methodologies and Applications*. 2nd edn, Wiley, Chichester.

Wilson, T.D. (1989) The Implementation of Information Systems Strategies in UK Companies: Aims and Barriers to Success. *International Journal of Information Management*, **9**, No. 4.

Wiseman, C. (1985) *Strategy and Computers*. Dow Jones Irwin.

Womack, J.P., Jones, D.T. and Roos, D. (1990) *The Machine That Changed the World*. Rawson Associates, New York.

Wright, R.N. (1988) Computer Integrated Construction. *Periodica* (Journal of the International Association for Bridge and Structural Engineering), **1**, February, pp. 17–24.

Yamazaki, Y. (1995) Computer Integrated Construction in a Japanese Construction Company. *Construction Management and Economics*, **13**, Special Issue on IT.

Yeo, K. (1991) Implementing a successful IT strategy for contracting firms. *International Journal of Project Management*, **9**, No. 1, 34–38.

Yetton, P., Johnston, K. and Craig, J. (1994) Computer Aided Architects: A Case Study of IT and Strategic Change. *Sloan Management Review*, **35**, No. 2, 57–67.

Zimmerman, C. (1990) In tight economy, information management is key to success. *International Spectrum*, January–February, 36–38.

Zuboff, S. (1988) *In the Age of the Smart Machine: The Future of Work and Power*. Heinemann Professional, Oxford.

Zuk, W. (1988) Structures beyond tomorrow. *Journal of Professional Issues in Engineering*, **114**, No. 3, 344–347.

Index